SOIL ENGINEERING

Testing, Design, and Remediation

SOIL ENGINEERING

Testing, Design, and Remediation

Fu Hua Chen, P.E.
Honorary Member, ASCE, 1999

Edited by
M.D. Morris, P.E.

CRC Press
Boca Raton London New York Washington, D.C.

Library of Congress Cataloging-in-Publication Data

Chen, F.H. (Fu Hua)
Soil engineering: testing, design, and remediation / Fu Hua Chen.
p. cm.
Includes bibliographical references and index.
ISBN 0-8493-2294-4 (alk. paper)
1. Soil mechanics. 2. Engineering geology. 3. Foundations. 4. Soil remediation. I. Title.
TA710.C5185 1999
624.1'51—dc21

99-23653
CIP

International Standard Book Number 0-8493-2294-4
Library of Congress Card Number 99-23653
Printed in the United States of America 1 2 3 4 5 6 7 8 9 0
Printed on acid-free paper

Foreword

A true Renaissance man, Fu Hua Chen was educated in both China and the United States. Returning to his homeland to contribute to its struggle against Japanese attrition, he was chief engineer on the Burma Road. That artery held together the victorious Allied campaign to end World War II on the Asian mainland.

After the Tibet Highway, the Ho Chi Minh Trail, and other large China projects, Dr. Chen brought his family to the U.S. to build a better life. Successful in that, he then devoted his remaining years to returning to his community, his society, and his profession some of the benefits American life had provided for him.

Acknowledged as the world's authority on expansive soils, Dr. Chen published books on that and other aspects of geotechnical engineering, and a riveting autobiography. He wanted the top rung of his career ladder to be his guide for constructors and consultants to demystify soils and foundation engineering. It is a plain-talk effort to help builders understand and deal with that complex facet so vital to construction.

With the publication of this book, Dr. Chen has achieved that goal, to top off a monumental career that ended peacefully among his family in his 87th year.

M.D. Morris, P.E.
Advisory Editor

Chen, Fu Hua
21 July 1912 — 5 March 1999
Civil Engineer, Author, Educator, Humanitarian

Introduction

When I was at the University of Michigan in 1935, I took a course on soils with Professor Hogentogler. He had just completed a book entitled *The Engineering Properties of Soil.* At that time, soil mechanics was not known. I talked to Dr. Terzaghi at Vienna in 1938; he assured me that he had nothing to do with the term "soil mechanics." We all realized that the term "mechanics" is associated with mathematics. By using the term "mechanics" with soil, the academicians firmly linked engineering with mathematics. It appears that in order to understand soil, one must understand "elasticity," "diffusion theory," "finite element" and other concepts. After several years of dealing with foundation investigation, most consultants realize that soil engineering is an art rather than a science as the academicians depicted.

In the last 40 years, no fewer than 50 books have been written on the subject of soil mechanics. Most of them were written for use in teaching. Only a few touched on practical applications. When engineers dealt with major complicated projects, such as the failure of the Teton Dam or the Leaning Tower of Pisa, high technology was required. However, 90% of the cases in which consulting engineers are involved do not require mathematical treatment or computer analysis; they mostly need experience. Consulting soil engineers are involved primarily with the design of foundation for warehouses, schools, medium-rise buildings, and residential houses. With such projects, the complete answers to soil engineering problems cannot be resolved solely with textbook information.

The purpose of this book is to provide consulting engineers with the practical meaning of the various aspects of soil mechanics; the use of unconfined compression test data; the meaning of consolidation tests; the practical value of lateral pressure; and other topics.

In addition to the technical aspect of foundation investigation, in the real world one should be aware that the shadow of litigation hangs over the consultant's head. A careless statement may cost the consultant a great deal of time and money to resolve the resulting legal involvement.

It is expected that the academicians may find many inconsistencies in this book. However, at the same time, I expect that the book will find its way to the consulting engineer's desk.

Acknowledgments

I wish to thank Professors Ralph Peck and George Sowers, geotechnical engineers whom I greatly respect, for their encouragement in preparing this book. I have quoted directly from their publications in many places.

I also wish to thank the American Consulting Engineers Council and the Association of Soil and Foundation Engineers for the benefit of using their publications.

The manuscript was edited and revised with many valuable suggestions from:

Paul Bartlett, Honorary Member, ASCE, Dean Emeritus, University of Colorado at Denver;
Richard Hepworth, P.E., President, Pawlark and Hepworth, Consulting Engineers;
M.D. Morris, P.E., F.ASCE, Ithaca, New York;
Dr. John Nelson, Professor, Colorado State University;
Malcolm L. Steinberg, P.E., F.ASCE, Steinberg & Associates, El Paso, Texas.

Dr. Jiang Lieu-Ching, University of Colorado at Denver, and Mr. Tom Jenkins, writer, also helped with many details.

To my wife Edna, with love and appreciation;
she took care of me during the preparation of this book while
I was suffering severely from emphysema.

Table of Contents

1 Site Investigation

CONTENTS

The stability and performance of a structure founded on soil depend on the subsoil conditions, ground surface features, type of construction, and sometimes the meteorological changes. Subsoil conditions can be explored by drilling and sampling, seismic surveying, excavation of test pits, and by the study of existing data.

Elaborate site investigation oftentimes cannot be conducted due to a limited assigned budget. For very favorable sites, such investigation may not be warranted. However, if the area is suspected of having deep fill, a high water table, or swelling soil problems, extensive soil investigation will be necessary even for minor structures. The soil engineers should not accept jobs in problem areas without thorough investigation. Bear in mind that in court of law, limited budgets or limited time frames are not excuses for inadequate investigation. Differing site conditions are a favorite tool of the contractors. They are used as the basis for extra claims on their contracts.

Since a consulting soil engineer cannot afford to treat each site as a potential hazard area, the amount of investigation required will generally be dictated by the judgment and experience of the engineers. If the project is completed on time and under budget, the consultant may still be criticized for being too conservative. On the other hand, if problems are encountered in the project, no number of excuses can relieve consultants of their responsibility.

1.1 GENERAL INFORMATION

The content of this chapter has very little to do with soil engineering. However, as a consultant, site investigation is probably one of the most important parts of the total inquiry or the report. Average owners know very little about engineering, but they do know a great deal about the property they own. Misrepresentation of the observations can often cause a great deal of trouble. For instance, describing the property as located in a low-lying area may devalue the property. Pointing out the cracks in the building owned by someone else in the neighborhood may induce the buyer to decrease the offer and in extreme cases may result in litigation.

Valuable information about the presence of fills and knowledge of any difficulties encountered during the building of other nearby structures may be obtained from talking to older residents of the area.

Much of the site investigation depends on the experience and good judgment of the field engineer or the technician. An experienced field engineer has the sense of a bloodhound; he is able to smell or sense a problem when he visits the site. A red flag will be raised to call for thorough investigation. In a potential swelling soil area, special attention should be paid to the condition and foundation system of the existing structures.

When the site is located out of town, consulting engineering firms sometimes assign site investigation to a technician or a field man, who has little geotechnical experience. He may ignore some important features which should be pointed out in the geotechnical report. An experienced technician with many years of training in a geotechnical company can be worth more than an engineer freshly out of college with a Ph.D. degree.

Generally, it is a small building with inadequate funding, poor planning, and a low-bidding contractor that presents the most trouble. The owner of such a project generally considers soil investigation as a requirement fulfillment rather than a protection against foundation failure. Geotechnical engineers should ask for more details regarding the site condition and proposed construction before accepting such assignments.

1.1.1 PROPERTY

In most cases, the owner's property is well defined. However, one often comes across property that is not surveyed and not clearly marked. It is quite possible that the field man located his test hole outside of the property line. There would be a great deal of argument on the liability of such an incident. It is not unusual that the engineering company has to pay for the damage. There are cases when the upper portion of the retaining wall is within the property line, but the base of the wall extends to the neighboring property. There are cases when the surveyor's monument is intentionally moved for the benefit of the owner. If the owner is on good terms with his neighbor, nothing will happen. Otherwise, the case may wind up in court, and the engineers may be involved.

Errors in property lines may lie undetected long after the project is completed and forgotten. The mistake may involve the demolition of the existing structure. It is also possible that the client did not acquire the final title to the property and moved ahead of schedule to order the soil test. The result in one case was the field engineer being chased by an angry owner with a shotgun.

After the property lines have been established, permission should be obtained from the owner to enter the property with drilling equipment. This should be in writing, although oral permission in front of a witness may be enough. The following is a summons filed by the owner:

> "…None of the defendants asked for or obtained plaintiff's permission to enter into and to explore his leasehold estates, did not ask plaintiff for permission to drill or have drilled a rotary hole into his leasehold, and did not ask plaintiff for permission to have geophysical and geological testing conducted pertaining to his leasehold…"

It is obvious that in this case the engineering company is liable.

1.1.2 Accessibility

Not all properties are accessible to drilling equipment. Oftentimes, the site is covered with crops. It is a sad sight to see crops ruined by a drilling vehicle. The engineering company, not the owner, will wind up paying for the damage.

In mountain sites, access usually presents a problem. Before sending the drilling equipment to the site, a general survey of the route to enter the site should be made. Sometimes, trespassing on the neighboring properties cannot be avoided. In such cases, permission should be obtained.

If the property is fenced, permission should be obtained to open the gate. Be very sure that the gates are properly closed after entering or leaving. The loss of cattle or prize horses certainly can add to the liability bill. In the eyes of the attorney, anything lost is not replaceable.

In soft ground, as at the time of spring thaw or after continuous rain, it is a lost effort to move the drilling equipment to the site. In order to avoid loss of time or the cost of towing, it is always advisable to evaluate the accessibility first. An all-terrain drill rig is able to move into places where conventional drill rigs cannot gain access. In this case, the client should agree to pay for the additional cost or wait until the ground has dried up.

During winter months, it is better to move the rig in the early morning when the ground is frozen and move out before thawing. Profit and loss on a project depend sometimes on the intelligent planning of the field engineer. Accessibility problems should be considered before a cost estimate is offered. The margin of profit for a consulting firm is very thin.

1.1.3 Records

A complete record of the site investigation should be maintained by the field engineer. This includes the time, date, the names of all parties involved, and all letters and notes. Such records appear to be so obvious and unnecessary at the time, but may turn out to be invaluable in a court of law at a later date. Dates are important in that conditions such as water tables and climates change with time.

Some field engineers are required to describe the site by filling out standard questionnaires. Such lengthy questionnaires may not be desirable. Most items listed in the questionnaires are unrelated and unnecessary, while vital issues can be neglected. A field engineer should treat each site as an individual case and use his observation and judgment in recording all pertinent details.

1.1.4 Utility Lines

Before sending the drill rig to the site, subsurface utility lines should be checked out thoroughly. Standard contracts between the consulting engineering company and the client usually specify that the company will not be responsible for subsurface structures not indicated on the plans furnished to the engineer. However, in the case of accident, the information furnished to the engineering company cannot protect the geotechnical engineer from being named as a defendant. For projects near a metropolitan area where the site is crisscrossed with utility lines in addition to those indicated in the existing plans, it is important to notify the telephone company, the public service company, the water works, and the city engineer on the project. The concerned parties will send agents to the site to accurately delineate the location of the various lines.

In one project at the Stapleton Airport in Denver, the engineer was provided with the location of the underground cable and all utility lines. The engineer did not check the date on the plot, which was made several years before. During drilling, a main fuel line was damaged, causing the delay of all air traffic. The incident was finally settled out of court. Luckily, the geotechnical company was a relatively new firm with few assets and was able to get away with limited payment.

In residential areas, the location of a sprinkler system should also be checked out. The chance of hitting a 1-in. utility line with a 4-in. auger in several acres of open field appears to be remote, but in fact such incidents have taken place over and over.

1.1.5 Existing Structures

The behavior of the existing structures has an important bearing on the selection of the proposed structure. All possible information should be obtained concerning structures at the site and in the immediate proximity. Inquiry should be made as to the condition of the structure, age, and type of foundation. If adjacent existing structures have experienced water seepage problems, the possibility of a high water table condition or a perched water condition in the area is likely to exist. The best way to determine such a condition is to enter the lower level of the building and look for watermarks on the wall.

It is not often that a geotechnical engineer has an opportunity to examine the cracking of the existing building located on or near the project site. By studying the condition of the existing structure, one will be able to tell the adequacy of the existing foundation system. If there are cracks in the foundation system, it is certainly important to try to determine the cause of the cracking. The cracking can be caused by foundation settlement, swelling of the foundation soils, or even from an earthquake. The age of the structure may provide the potential for distress. An experienced geotechnical engineer treats the cracking as if it is the writing on the wall.

If the existing building is in excellent condition, this does not mean that the existing building system can be used for the design of the new structure. The existing structure's foundation system could have been overdesigned. This is especially true in the case of old structures where massive foundation systems were traditionally used.

1.1.6 Additions

A portion of a geotechnical consultant's project is in addition to the existing structures. Building owners may not want to use the initial consultant and will approach a new consultant for the geotechnical study for one of the following reasons:

The initial geotechnical firm is not available.
The fee charged by the initial engineering company is too high.
The initial recommendation of the foundation system cost is too high.
The possibility of using another foundation system.
The initial building suffered damage.

If the cost of consultation is the main reason, consideration should be given to rejecting the job. This is on account of breaching the ethical practice. If the structure suffered damage, the field engineer should determine as closely as possible the following:

Damage caused by using the wrong foundation system
Damage caused by reasons other than soil
Damage caused by poor maintenance

Bearing in mind that the geotechnical engineer cannot guarantee the performance of the structure, the second consultant should be prepared to defend the initial consultant in a court of law rather than condemn him. It is a mistake to brag about one's knowledge by pointing a finger at one's fellow engineer.

Realizing the importance of site conditions to a geotechnical consultant and the responsibility the engineer is confronting, the Associated Soil and Foundation Engineers (ASFE) proposed an agreement between the owner and the engineers as follows:

1. The owner shall indicate to the soil engineer the property lines and is responsible for the accuracy of markers.
2. The owner shall provide free access to the site for all necessary equipment and personnel.
3. The owner shall take steps to see that the property is protected, inside and out, including all landscaping, shrubs, and flowers. The soil engineer will

not be responsible for damage to lawns, shrubs, landscapes, walks, sprinkler systems, or underground utilities and installations caused by movement of earth or equipment.

4. The owner shall locate for the soil engineer and shall assume responsibility for the accuracy of his representations as to the underground utilities and installations.

Such an agreement when signed should be sufficient to protect the engineering company, yet a talented lawyer may still find loopholes that involve the engineer.

At the same time, a large portion of geotechnical investigation is carried out without a written contract. Small projects are carried out based on a single-page letter or even oral agreement. In such cases, the roles of the field engineer become more and more important. Unfortunately, the importance of the field engineer is seldom realized by the consulting firm until a summons is served.

1.2 TOPOGRAPHY, GEOLOGY, HYDROLOGY, AND GEOMANCY

Topography, geology, and hydrology should be treated as an integral part of soil engineering. No soil engineer can be considered knowledgeable if he lacks information on these subjects. No soil report can be considered complete without touching on these subjects. No investigation can be considered satisfactory without having such subjects in mind.

Such information can be obtained by reviewing available data, studying existing maps, or making a reconnaissance survey. Care must be taken as to the accuracy of such information. Oftentimes, site grading can completely alter the topography, and development in the neighborhood can alter the hydraulic balance.

1.2.1 TOPOGRAPHY

Topography is defined as the features of a plain or region. Generally, for larger projects a topographic survey is available. Care must be taken with the date of the survey and the bench mark referred to. Sometimes site grading can completely alter the original ground features. Topography can be different if the original photogrammetric survey was taken when the site was vegetated.

Outdated contour elevation should not be used for elevation of the top of the drill holes without careful checking.

The shape of gullies and ravines reflects soil textures. Gullies in sand tend to be V-shaped with uniform straight slopes. Gullies in silty soils often have U-shaped cross-sections. Small gullies in clay often are U-shaped, while deeper ones are broadly rounded at the tops of the slopes.

The location of natural and man-made drainage features is also of importance. Erecting a structure across a natural gully always poses a future drainage problem. The water level in any nearby streams and ponds should be measured and recorded.

Irrigation ditches can be dry during most of the year but can carry a large amount of water during irrigation season. Water leaking out from the ditches and ponds can supply moisture to the foundation soil and cause settlement or heaving of footing and slab. The lifting of drilled piers in an expansive soil area due to the infiltration of water from ditches and ponds is not uncommon. The source of water may not be detected for a long time.

Water leaking out from the ditches can also cause the cracking and dampness of the basement slab. The location and elevation of the ditches should be included as part of the engineering report.

Streams and nearby runoffs are important parts of the investigation. Engineers should pay special attention to the extent of the flood plain. Such preliminary information can usually be obtained from the U.S. Geological Survey, or the U.S. Department of Agriculture Soil Survey reports.

The steepness of valley slopes is of special concern for sites chosen in mountain areas. Some environmental agents classify valley slopes in excess of 30° as potential hazard areas. Slope stability depends upon the slope's angle, rock and soil formations, evidence of past slope movement, and drainage features. The field engineer should be aware of the possible slope problems associated with landslides, local slope failure, mud flow, and other problems. The vegetative cover on the slope, shapes of tree, and the behavior of any neighboring structures should also be known.

1.2.2 Geology

Geology is the science of the earth's history, composition, and structure. Branches include mineralogy, petrology, geomorphology, geochemistry, geophysics, sedimentation, structural geology, economic geology, and engineering geology. The last category is of utmost concern to the foundation engineers.

The science of geology existed long before the advance of soil engineering. Colleges offer geology to most civil engineering students. However, some professors in geology may have little knowledge of engineering. Consequently, the relationship between geology and soil mechanics is seldom stressed. Students do not pay much attention to geology and give such courses the same weight as astronomy or chemistry.

It is not until an engineer enters the field of consulting that he realizes the close relationships between soil mechanics and geology. For average small structures that do not require special foundation designs, geology information may not be required. It is a mistake for consultants to put a section in their reports on geology if the content has no bearing on the project. A section on geology in the consultant's report is necessary only when such information is vital to the project. Consulting firms should have qualified staff geologists to conduct and study such projects. If the soil engineer is not a qualified geologist, he or she should not attempt to touch the subject.

A geological assessment based on prior knowledge of the area may be required before the study can be completed. The geological assessment can describe any geological conditions that have to be considered before any soil testing is initiated and recommendations for the foundation design are presented.

General surficial geology of the area includes the study of slopes, tributary valleys, landslides, springs and seeps, sinkholes, exposed rock sections, origin of deposit, and the nature of the unconsolidated overburden.

An inspection of upland and valley slopes may provide clues to the thickness and sequence of formations and rock structure. The shape and character of channels and the nature of the soil may provide evidence of past geologic activity. An engineering geologist should identify and describe all geologic formations visible at the surface and note their topographic positions. The local dip and strike of the formations should be determined and notes made of any stratigraphic relationships or structural features that may cause problems of seepage, excessive water loss, or slide of embankment.

Some of the geological concerns to the foundation engineers are as follows:

The bearing capacity of bedrock
The bearing capacity and settlement of windblown deposits
The expansion potential of shale
The orientation of the rock formation
The excavation difficulty
The drilling problem
The slope stability

In the mountain areas, the mapping of surficial geologic features is highly desirable. Features to be shown on the map should include:

Texture of surficial deposits
Structure of bedrock, including dip and strike, faults or fissures, stratification, porosity and permeability, schistosity, and weathered zones
Area of accelerated erosion deposit
Unstable slopes, slips, and landslides
Fault zones

Geologists as well as experienced engineers should be able to recognize a potential swelling soil problem. For instance, the red siltstone formation in Laramie, Wyoming will not pose a swelling problem, while a few miles to the west where claystone of Pierre formation is observed, swelling can be critical. In the front range area west of the foothills, claystone shale dips as much as 30° with the horizontal. The joints within the rock can allow easy access of water and cause volume change. Such problems should be carefully studied.

To assure adequate planned development of a subdivision, some state laws require the subdivider to submit such items as:

Reports concerning streams, lakes, topography, geology, soils, and vegetation
Reports concerning geologic characteristics of the area that would significantly affect the land use and the determination of the impact of such characteristics on the proposed subdivision
Maps and tables concerning suitability of types of soil in the proposed subdivision

Careful discussion between geotechnical engineers and geologists should be maintained during the writing of the report. The report should be presented as a whole. It should give the client the impression that the report is from one author. Contradictory opinions between geologists and geotechnical engineers should be settled before the report is completed.

1.2.3 Hydrology

Hydrology is defined as the scientific study of the properties, distribution, and effects of water on the earth's land surface in the soil and rock. Geotechnical engineers are dealing with water all the time. As Terzaghi stated, "without any water there would be no use for soil mechanics." The most common issues a geotechnical engineer encounters are permeability, seepage, and flow in connection with ground water. For major projects such as dams and canals, the geotechnical engineer should seek advice and consultation from a hydrologist.

The permeability property of soil cannot be separated from the soil drainage characteristics. The former has been researched both in theory and in the laboratory by the academicians, yet the design and construction of the drainage installation seldom receive proper attention from the architect. Long drainage facilities are often shown on the design drawing by mere dotted lines. Construction of the drain facilities is often left in the hands of the builders. It is not uncommon to see that the drains were constructed with reverse grade or without proper outlet. Since drainage facilities are generally installed below ground surface, the defective systems are seldom revealed.

Details of drainage and soil moisture control will be discussed in the subsequent chapters.

1.2.4 Geomancy

Geomancy, or as what the Chinese refer to as "feng shui" or "Wind and Water" is defined as the art of adapting the residence of the living and the dead so as to harmonize with the cosmic breath.

The ups and downs of the profile of the land is of vital importance to the quality of the site. The ground must be hard and solid and must have a good profile like that of the real dragons if it is rated as a good site. The sand on the ground and the water sources are also of great importance to the geomancer. From a geotechnical point of view, the ideal site would be one with hard bedrock overlain by granular deposits.

In fact, no family or business will consider building on a piece of land without the consultation with a geomancer.

In the Western world, feng shui has been considered dogmatic faith or superstition. However, this art has been under intense study in recent years in the U.S. as well as in European countries. It is not surprising that one will find a link between feng shui and geotechnical engineering.

REFERENCES

J. Atkinson, *An Introduction to the Mechanics of Soils and Foundations,* McGraw-Hill, New York, 1993.

L. Evelyn, *Chinese Geomancy,* Times Books International, Singapore, 1979.

G.B. Sowers and G.S. Sowers, *Introductory Soil Mechanics and Foundations,* Collier-Macmillan, London, 1970.

K. Terzaghi, R. Peck, and G. Mesri, *Soil Mechanics in Engineering Practice,* John Wiley-Interscience Publication, John Wiley & Sons, New York, 1995.

2 Subsoil Exploration

CONTENTS

Subsoil exploration is the first step in the designing of a foundation system. It consists essentially of drilling and sampling. The process of subsoil exploration took place long before soil mechanics was born. Present-day engineering requires thousands of exploratory test borings to build a structure like the Great Wall of China. The wall actually winds around the mountains, avoiding problem soil areas. Somehow its ancient builders had a sense in selecting the good foundation soils.

Chinese legend tells the story of a commandeered laborer who died while building the Great Wall. His wife's lament at the foot of the wall was so moving that the wall collapsed. We suspect now if the story is true, the wall collapsed due to foundation failure.

Experienced engineers use soil mechanics to *confirm* their conclusions rather than to reach their conclusions. Many factors affect the choice of a subsoil exploration program; the judgment of the engineer is deemed necessary.

2.1 DIRECT METHODS

The most suitable method to perform subsoil exploration depends on the type of soil in the general area; type of equipment available; ground water condition; type of proposed structure; and the amount of money allocated for the exploration.

Direct methods of exploration include the digging of test pits and the use of auger drilling or rotary drilling. Each method has its merits and its drawbacks. The engineer must use his or her judgment based on experience and the evaluation of the site conditions to select the best method. Unfortunately, in most projects there is very little choice. The driller and the available equipment are the only choices. The engineer often must help the driller in solving the drilling problems.

2.1.1 Test Pits

Probably the most accurate subsoil investigation method is the opening of test pits. In a test pit, the engineer can examine in detail the subsoil strata, stratification, layer and lens, as well as take samples at the desired location. However, the use of test pits is limited by the following:

When the depth of the test is limited to the reach of a backhoe, generally 12 ft.
When the investigation involves basement construction that extends below the ground level
When the water table is high, which prevents excavation
When the soil is unstable and has the tendency to collapse, this prevents the engineer from entering the pit. Entering a test pit can involve certain risks and the regulations of the Occupational Safety and Health Administration (OSHA) should be observed
When the standard penetration resistance test is required

In locations where subsoil consists essentially of large boulders and cobbles, the use of test pit investigation is most favorable. Auger drilling through boulders and cobbles is difficult. The cost of rotary drilling may not be warranted for small projects.

The layman's conception of subsoil investigation generally assumes that drilling to a great depth constitutes the main portion of the cost. After drilling, the layman thinks the remaining task of the engineer, such as testing and preparation of the report, is of minor importance. Consequently, when no drill rig shows up at the project site, the client feels that he has been cheated and the money paid for the investigation is not justified. With such a philosophy, the engineering company usually attempts to drill each project when possible instead of resorting to the use of a backhoe.

Another possible investigation method is to drill a large-diameter caisson hole to the required depth; a caisson rig is shown in Figure 2.1. By entering the hole, the engineer can clearly examine the subsoil strata and undisturbed samples can be obtained at the desired depth. However, such practice is frowned upon by OSHA for safety reasons.

If deep excavation is required or if the soil cannot maintain the steep side slope, sometimes the use of sheeting is necessary. The minimum size of the pit is 4 × 4 ft, so that a man can enter the pit. Pit excavation in this case must be carried out entirely by manpower. The cost of such an operation is high.

2.1.2 Auger Drilling

In auger drilling, the hole is advanced by rotating a soil auger while pressing it into the soil, and later withdrawing and emptying the soil-laden auger. A series of augers

FIGURE 2.1 Caisson rig with a rock bit.

and a special drilling machine for their operation have been developed by many specialized soil exploration equipment companies.

A series of augers, or a continuous flight helical auger (Figure 2.2), are used for drilling holes with diameters of 4 to 8 in. to a depth of 100 ft. An auger boring is made by turning the auger the desired distance into the soil, withdrawing it, and removing the soil for examination and sampling. As the depth increases, new auger sections are added. It is difficult to determine the depth from which the soil discharged from the auger is excavated. Consequently, in order to obtain a representative sample or an undisturbed sample, it will be necessary to stop the drilling and replace the auger with a sampler. The sampler can then be pushed or driven into the soil at the desired depth.

Auger drilling can be successfully conducted in almost all types of soils and in shale bedrock. For hard bedock such as limestone, sandstone, and granite, rotary drilling is necessary.

The drilling machine has a folding mast with a chain-operated feed and lifts. In wet and spongy terrain where a tire-mounted rig is not accessible, a tractor-mounted type rig is available (Figure 2.3). More than 90% of soil exploration study today is conducted by such a device or similar devices.

Relatively short helical augers with interchangeable cutters are used for medium-size holes, whereas large-diameter holes up to 24 in. are excavated by means of a disc auger.

FIGURE 2.2 Continuous flight hollow stem auger.

FIGURE 2.3 Tractor-mounted auger drilling rig.

FIGURE 2.4 Diamond drill bit.

2.1.3 Rotary Drilling

In rotary drilling the bore hole is advanced by rapid rotation of the drilling bit, which cuts and grinds the material at the bottom of the hole into small particles. The cuttings are removed by pumping drilling fluid from a sump down through the drill rods and bit, up through the hole from which it flows first into a settling pond, and then back to the main pit.

Rotary drilling with a diamond bit (Figure 2.4) can be used efficiently for drilling through semi-hard rocks. Most truck-mounted drill rigs (Figure 2.5) can be used for rotary drilling with little modification. Core samples are brought up by the drill and can be visually examined. The general characteristics, particularly the percentages of recovery, are of importance to foundation design and construction.

For consultants with limited experience with drilling, it is better to study catalogs from various companies before deciding on the type of equipment best suited for the job. An experienced operator is essential for handling and maintaining the costly equipment.

2.2 INDIRECT METHODS

The geophysical method of exploration is the main indirect method of subsoil exploration.

In subsoil investigation, the seismic method is most frequently used. Seismic methods are based on the variation of the wave velocity in different earth materials. The method involves generating a sound wave in the rock or soil, using a sledge-hammer, a falling weight, or a small explosive charge, and then recording its reception

FIGURE 2.5 Truck-mounted drilling rig.

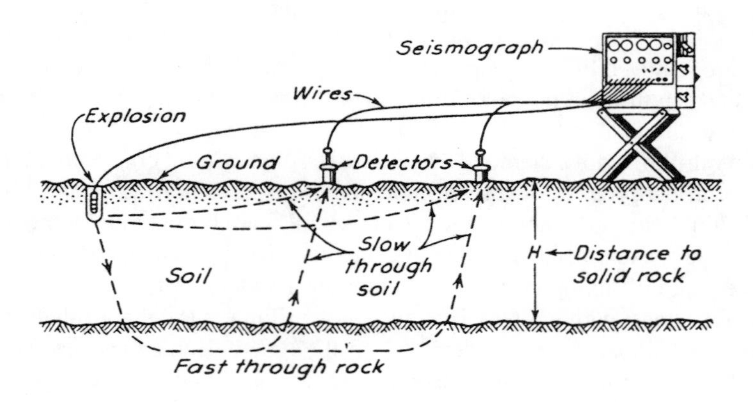

FIGURE 2.6 Diagram of seismic refraction test (after Moore).

at a series of geophones located at various distances from the shot point, as shown in Figure 2.6. The time of the refracted sound arrival at each geophone is noted from a continuous reader. Typical seismic velocities of earth materials in ft/sec are shown in Table 2.1.

TABLE 2.1
Seismic Velocities of the Wave Velocity in Different Earth Materials

Material	Velocity
Dry silt, silt, loose gravel. loose rocks, talus, and moist fine-grained soil	500–600
Compacted till, indurated clays, gravel below water table compacted clayey gravel, cemented sand, and sandy clay	2500–7500
Rock, weathered, fractured, or partly decomposed	2000–10,000
Sandstone, sound	5000–14,000
Limestone, chalk, sound	6000–20,000
Igneous rock, sound	1200–20,000
Metamorphic rock, sound	1000–16,000

(after Peck, Hanson, and Thornburn)

Seismic exploration requires the following:

1. Equipment to produce an elastic wave, such as a sledgehammer used to strike a plate on the surface;
2. A series of detectors, or geophones, spaced at intervals along a line from the point of origin of the wave;
3. A time-recording mechanism to record the time of origin of the wave and the time of arrival at each detector.

The setup is shown in Figure 2.6.

The advantage of seismic exploration is that it permits a rapid coverage of large areas at a relatively small cost. The method also is not hampered by boulders and cobbles which obstruct borings. Seismic survey is frequently used in regions not accessible to boring equipment, such as the middle of a rapid river.

The disadvantage of the exploration method is the lack of unique interpretation. It is particularly serious when the strata are not uniform in thickness nor horizontal. Irregular contacts often are not identified and the strata of similar geophysical properties sometimes have greatly different properties.

Whenever possible, seismic data should be verified by one or two borings before definite conclusions can be reached. As advice to young engineers, seismic data when used in a geotechnical report should be qualified as to possible error and the margin of error.

2.3 TEST HOLES

The number of the test holes and the depth required for a project depend on the type of foundation system, uniformity of the subsoil condition, and to some extent the importance of the structure.

2.3.1 Test Hole Spacing

If preliminary investigation indicates that shallow footings will be the most likely type of foundation, then it will be desirable to have the drill holes closely spaced to better evaluate the subsoil condition within the loaded depth of the footings. If, however, a shallow foundation system is not feasible and a deep foundation is likely to be required, then the number of test holes can be decreased and the spacing increased.

As a rule of thumb, test holes should be spaced at a distance of 50 to 100 ft. In no case should the test holes be spaced more than 100 ft apart in an expansive soil area or in a problem soil area.

It is a common misconception that drilling of test holes is the major cost of the subsoil investigation. Consequently, there is the tendency to drill as few holes as possible, preferably only one. The risk involved in such an undertaking is enormous. Erratic subsoil conditions can exist between widely spaced test holes.

For the foundation system of a high-rise building in Chicago, a renowned engineer drilled only one test hole, which was the owner's requirement. For a complete investigation, at least four test holes were required. Fortunately, the subsoil within the building area was unusually uniform, and nothing happened to the building.

Closely spaced test holes are especially important where the presence of expansive soils is suspected. For instance, if sandstone and siltstone bedrock are present at a shallow depth, a logical recommendation is to found the structure directly on the bedrock, with spread footing designed for high pressure.

Geotechnical engineers in the Rocky Mountain states, when conducting foundation investigation in a subdivision, insist on drilling one test hole for each lot. This is to ensure that the swelling potential and water table conditions are adequately covered.

Engineers should use special care when dealing with a site containing man-made fill. Since man-made fill is placed at random, it is not possible to delineate the extent of the fill. Field engineers should locate the test holes based on their experience and judgment. Such conditions should be clearly documented in the soil report to avoid possible legal problems.

In one commercial project, the engineer drilled one test hole at each corner on the site of the proposed structure and found the subsoil very uniform. No further drilling appeared to be necessary. During construction, the owner found deep garbage fills at the middle of the site. Such a finding not only required further drilling but also upset the entire recommendations given in the report.

2.3.2 Test Hole Depth

The depth of the test hole required is generally governed by the type of foundation. Before drilling, the field engineer has little conception of the probable foundation system. Therefore, the first hole drilled should be deep enough to provide information pertinent to both shallow and deep foundation systems. Samples in the hole should be taken at frequent intervals, preferably not more than 5 ft apart. After the completion of the deep test hole, the field engineer should have a fairly good idea of the possible foundation system. Consequently, for the subsequent holes, more attention should be directed to the upper soils if a shallow foundation system is likely. If, however, a deep foundation system is contemplated, drilling to bedrock would be necessary for all holes.

Often, the depth to bedrock is the criterion as to the depth of all test holes. Where bedrock is within economical reach, say within 40 ft, it is advisable to drill a few holes into bedrock, irrespective of the foundation system. In the Rocky Mountain region, the depth to the top of the shale is extremely erratic, and the depth to bedrock can increase as much as 30 ft within a short distance. This should be taken into consideration when determining the location and depth of the test holes.

In some cases, deep holes are required, not for the foundation system requirement but for the determination of the water table elevation. For deep basement construction, the depth of the test holes should be at least 20 ft to preclude the possibility of groundwater becoming a problem on the lower floor.

At times, the architect or the structural engineer wants to dictate the location and depth of the test holes. They want to do that as they think the consulting fee for the geotechnical engineers can be better controlled. An established firm should not accept such a restriction. The geotechnical engineers should have a free hand in planning the investigation.

2.3.3 Water in the Test Hole

The depth of the water table as measured during drilling should be carefully evaluated. It is always necessary to wait for at least 24 hours to check on the stabilized water table for the final measurement. Property owners sometimes will not allow open drill holes. The technician should plug the top of the drill holes and flag them for identification.

If the water level in the drill holes is allowed to drop below ground level, when the drill rods are removed rapidly, an upward hydraulic gradient is created in the sand below the drill hole. Consequently, the sands may become quick and the relative density may be greatly reduced. The penetration resistance value will be accordingly much lower than that corresponding to the relative density of the undisturbed sand. Care is required to ensure that the water level in the drill hole is always maintained. Any sudden drop or rise of the water table or a sudden change in the penetration resistance should be carefully recorded in the field log.

2.4 SAMPLING

The purpose of drilling test holes is not only for the observation of the subsoil conditions but also for obtaining representative samples. Both disturbed and undisturbed samples are valuable to geotechnical engineers. The undisturbed samples can be used for the determination of the stress strain characteristics of the material. Certain amounts of disturbance during sampling must be regarded as inevitable.

After accepting the foundation investigation assignment the geotechnical consultant should draft a program of field investigation for the field engineer to follow. The instruction should consist of the frequency and spacing of the test holes, the depth of the test holes, the field test required, etc. The field engineer should use his or her own judgment to determine whether the instruction should be modified. It is important that the field engineer not leave the site until all the information is gathered. The consultant cannot usually afford to investigate the site twice. Unlike some government projects where cost overruns can be tolerated, the consulting business is highly competitive; undue expense generally results in financial loss.

2.4.1 Disturbed Samples

Disturbed samples can be collected during the drilling process. Sometimes they can be collected without interrupting the operation. Samples can be collected from the auger cuttings at intervals. The field engineer should be sure that soils from different strata will not become mixed during drilling. Samples collected must represent soils from each different stratum. Disturbed samples can be stored in fruit jars. They should be sealed to retain the in situ moisture content and properly labeled.

In test pit excavation, large samples will sometimes be required in order to fulfill the laboratory testing requirements. Such samples should be at least 12 × 12 in. in size, wrapped in wax paper, and carefully transported to the laboratory. After it is carefully trimmed to the desired size, such a sample can be considered as undisturbed.

Representative samples can usually be obtained by driving into the ground an open-ended cylinder known as "Split Spoon." Spoons with an inside diameter of about 2 in. consist of 4 parts: a cutting shoe at the bottom; a barrel consisting of a length of pipe split into one half; and a coupling at the top for connection to the drill rod.

2.4.2 Undisturbed Samples

A very simple sampler consists of a section of thin-walled "Shelby" or seamless steel tubing which is attached to an adapter, as shown in Figure 2.7. The adapter or the sampler head contains a check valve and vents for the escape of air or water. A sample can be obtained by pushing the sampler into the soil at the desired depth. The operation must be performed carefully so as to experience minimum deformation. The principal advantages of the Shelby tube sampler are its simplicity and the minimal disturbance of soil.

A modification in the design of the split spoon sampler allows the insertion of brass thin-wall liners into the barrel. Four sections of brass liners (each 4 in. long) are used. Such a device allows the sampling and penetration test at the same time. This method was initiated in California by Woodward, Clyde and Associates and is known as the "California" sampler. It has been adopted throughout the Western U.S.

Samples of rock are generally obtained by rotary core drilling. Diamond core drilling is primarily in medium-hard to hard rock. Special diamond core barrels up to 8 in. in diameter are occasionally used and still larger ones have been built.

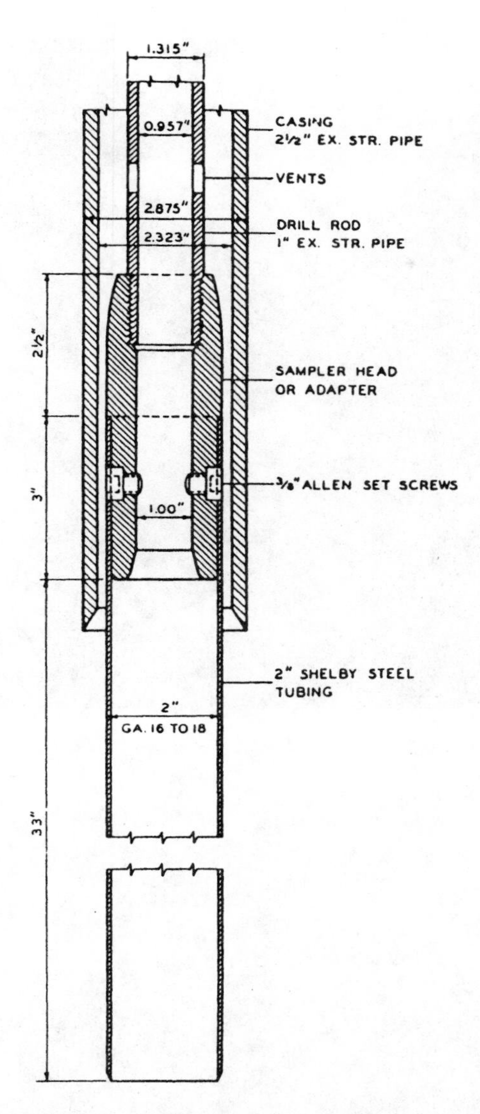

FIGURE 2.7 Shelby Tubing Sampler (after Moore).

In China, during the foundation investigation of the world's largest dam, the Three Gorges Dam, a special coring machine was used. The cores were up to 42 in. diameters and were taken at depths of about 200 ft deep, as shown in Figure 2.8. Such large samples enable the geologist to study the formation and texture of the foundation rock in detail.

FIGURE 2.8 Large diameter rock samples at the Three Gorges Dam, China.

REFERENCES

R. Peck, W. Hanson, and T.H. Thornburn, *Foundation Engineering*, John Wiley & Sons, New York, 1974.

K. Terzaghi, R. Peck and G. Mesri, *Soil Mechanics in Engineering Practice,* John Wiley-Interscience Publication, John Wiley & Sons, New York, 1995.

U.S. Department of the Interior, Bureau of Reclamation, *Soil Manual,* Washington, D.C., 1974.

R. Whitlow, *Basic Soil Mechanics,* Longman Scientific & Technical, Burnt Mill, Harrow, U.K., 1995.

3 Field Tests

CONTENTS

Field observation includes various numbers of tests. For building structures, the most commonly used tests involve the penetration resistance test, the drilling of test holes, and the opening of test pits. For hydraulic structure investigation, tests such as the permeability test, vane shear test, and others can be performed. Pavement and runway tests rely more on samples from core cutters, the California bearing ratio test, and others.

In recent years, unsaturated soils, including swelling and collapsing soils, have received a great deal of attention from geotechnical engineers. The performances of such soils are covered by specialized books and will be discussed only briefly in the following chapters.

Consulting engineers pay more attention to field test data than laboratory test results. Unfortunately, the engineer in charge cannot visit all sites, especially out-of-town projects. He must rely on his field engineer to perform all the necessary tests. All data collected from the field should be reviewed. All records should be checked for accuracy — but bear in mind that such documents may be brought up years later, when all the persons involved are no longer available.

3.1 FIELD TESTS FOR FOUNDATION DESIGN

Field investigation for foundation recommendations involves numerous tests. In situ testing includes the core cutter test, sand replacement test, standard penetration test, cone penetration test, vane shear test, plate bearing test, pressuremeter test, and many others. It is obvious that for a certain project not all tests are necessary. For shallow foundations, in situ testing is relatively easy, but for deep foundations such as piles and piers, field tests are often expensive and not always reliable.

3.1.1 Penetration Resistance Test

Probably the oldest method of testing soil is the "Penetration Resistance Test." In performing the Penetration Resistance Test, the split spoon sampler used to take soil samples is utilized. The split spoon is driven into the ground by means of a 140-lb hammer falling a free height of 30 in. The number of blows **N** necessary to produce a penetration of 12 in. is regarded as the penetration resistance. To avoid seating errors, the blows for the first 6 in. of penetration are not taken into account; those required to increase the penetration for 12 in. constitute the **N** value, also commonly known as the "blow count." The following should be considered in performing the penetration test:

1. **Depth Factor** — The value of **N** in cohesionless soils is influenced to some extent by the depth at which the test is made. This is because of the greater confinement caused by the increasing overburden pressure. In the design of spread footings on sand, a correction of penetration resistance value is not explicitly required. In other problems, particularly those concerned with the liquefaction of sand, however, a correction is necessary.
2. **Water Table** — When penetration is carried out below the water table in fine sands or silty sands, the pore pressure tends to be reduced in the vicinity of the sampler, resulting in a transient decrease in **N** value.
3. **Driving Condition** — The most significant factor affecting the penetration resistance value is the driving condition. It is essential that the driving condition should not be abused. The standard penetration barrel should not be packed by overdriving since, at this force, the soil acts against the sides of the barrel and causes incorrect readings. An increase in blow count by as much as 50% can sometimes be caused by a packed barrel.
4. **Cobble Effect** — The barrel will bounce when driving on cobbles; hence, no useful value can be obtained. Sometimes, a small piece of gravel will jam the barrel, thereby preventing the entrance of soil into the barrel, thus substantially increasing the blow count.
5. **California Sampler** — Considerable economy can be achieved by combining the penetration test with sampling as described under "undisturbed sample." Field tests have been conducted comparing the results of the penetration resistance of the California sampler with those of standard penetration tests. The tests indicate that the results are commensurable, with the exception of very soft soil ($\mathbf{N} < 4$) and very stiff or dense soil ($\mathbf{N} > 30$). By combining the penetration resistance test with sampling, more tests can be made and undisturbed samples can be obtained without resorting to the use of Shelby tubes.

With the exception of the area of saturated fine loose sands, the depth factor and the water table elevation factor can be disregarded. The results of the standard penetration test can usually be used for the direct correlation with the pertinent physical properties of the soil, as shown in Table 3.1.

TABLE 3.1
Penetration Resistance and Soil Properties of the Standard Penetration Test

Sands (Fairly Reliable)		Clays (Rather Reliable)	
Number of blows per ft, N	Relative Density	Number of blows per ft, N	Consistency
		Below 2	Very soft
0–4	Very loose	2–4	Soft
4–10	Loose	4–8	Medium
10–30	Medium	8–15	Stiff
30–50	Dense	15–30	Very stiff
Over 50	Very dense	Over 30	Hard

(after Peck)

The correlation for clay as indicated can be regarded as no more than a crude approximation, but that for sands is reliable enough to permit the use of **N**-value in foundation design. The use of **N** value below 4 and over 50 for design purposes is not desirable, unless supplemented by other tests. Some elaborate pile-driving formulae are based on field penetration resistance value. They should be used with caution, as the error involved in **N** value can be more than any of the other variables.

When driving on hard bedrock or semi-hard bedrock such as shale, if the amount of penetration is only a few inches instead of the full 12 in., it is customary to multiply the value by a factor to obtain the required 12 in.. For instance, if after 30 blows the penetration is only two inches, it is assumed that the **N** value is 120. Such an assumed value when used for the design of the bearing capacity of bedrock might be in error. An alternative is the pressuremeter test as described below, which may offer a better answer.

Some contracts call for a penetration test for every 5 ft and sampling at the same interval or every change of soil stratum. This may not be necessary. The field engineer should use his or her judgment to guide the frequency of sampling and avoid unnecessary sampling so that the cost of investigation can be held to a minimum. Samples in the upper 10 or 15 ft are important, as this is generally the bearing stratum of shallow footings. Soil characteristics at this level also govern the slab-on-grade construction and earth-retaining structures. Sampling and penetration tests at lower depths become critical when a deep foundation system is required.

3.1.2 Pressuremeter Test

The pressuremeter test is a device developed by Menard in 1950 for the purpose of measuring in situ strength and compressibility.

The probe is made up of a measuring cell, with a guard cell above and below, enclosed in a rubber membrane. The membrane is inflated, using water under an

FIGURE 3.1 Drilling for pressuremeter test.

applied carbon dioxide pressure. The pressure and volume readings are taken continuously. The two guard cells ensure that a purely radial pressure is set up on the sides of the bore hole. A pressure/volume-change curve is then plotted, from which shear strength and strain characteristics may be evaluated.

The pressuremeter test (Figures 3.1 to 3.3) can be used to evaluate the bearing capacity of shale bedrock at the bottom of large-diameter deep caissons.

3.1.3 Cone Penetration Test

The cone penetration test (Figure 3.4) is a static penetration test in which the cone is pushed rather than driven into the soil. The cone has an apex angle of 60° with a base area of 10 cm^2 attached to the bottom of a rod and protected by a casing. The cone is pushed by the rod at the rate of two cm/sec. The cone resistance is the force required to advance the cone, divided by the base area. The arrangement is known as the "Dutch Cone."

When the tip incorporates a friction sleeve, the base has an area of 15 cm^2. The local side friction is then measured as the frictional resistance per unit area on the friction sleeve.

The results of cone penetration tests appear to be most reliable for sand and silt that are not completely saturated. The application of the cone penetration test on stiff clay is limited.

3.1.4 Plate Bearing Test

The object of the plate bearing test is to obtain a load/settlement curve (Figures 3.5 to 3.7). For soil with relatively high bearing capacity, the load required to complete

FIGURE 3.2 Pressuremeter test.

the curve is often exceedingly high, and the cost of such testing is often unjustified. However, under certain circumstances where other test procedures are difficult to apply, such a test may be justified; for example, on weathered rocks, chalk, or hard-core fills.

The plate bearing test assures the client that the geotechnical engineer has taken the project seriously, and the recommendations presented are without errors. If the client is willing to pay for such a test just for assurance that nothing will go wrong, then the geotechnical engineer should be happy to comply with the client's wish, although the test results will not alter the recommendations in the report.

A pit is excavated to the required depth, the bottom leveled, and a steel plate set firmly on the soil. A static load is then applied to the plate in a series of increments, and the amount and rate of settlement measured. Loading is continued until the soil under the plate yields. A number of tests will be required using different plate diameters at different depths.

3.2 FIELD TESTS FOR HYDRAULIC STRUCTURES

In addition to the standard penetration resistance test and in-place density test for hydraulic structures, such as dams and canals, the field permeability test and the vane test are often performed. It is understood that for major dam projects, elaborate testing should be performed before and during construction. Approximate values of

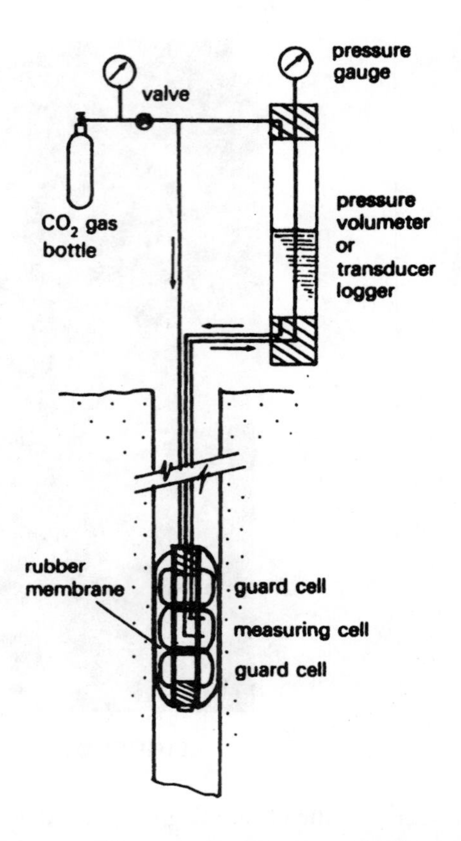

FIGURE 3.3 Menard Pressuremeter (after Whitlow).

permeability of individual strata penetrated by borings can be obtained by making water tests in the holes.

3.2.1 Open End Test

The test is made in an open end cased bore hole. After the hole is cleaned to the proper depth, the test is begun by adding water through a metering system to maintain gravity flow at a constant head. A surging of the level within a few tenths of a foot at a constant rate of flow for about 5 min is considered satisfactory. The permeability is determined from the following relationships:

$$k = Q/5.5\ r\ H$$

Where: k = permeability,
Q = constant rate of flow into the hole (gallons per minute),
r = internal radius of casing (feet), and
H = differential head of water (feet).

If it is necessary to apply pressure to the water entering the hole, the pressure in the unit of head is added to the gravity head as shown in Figure 3.8.

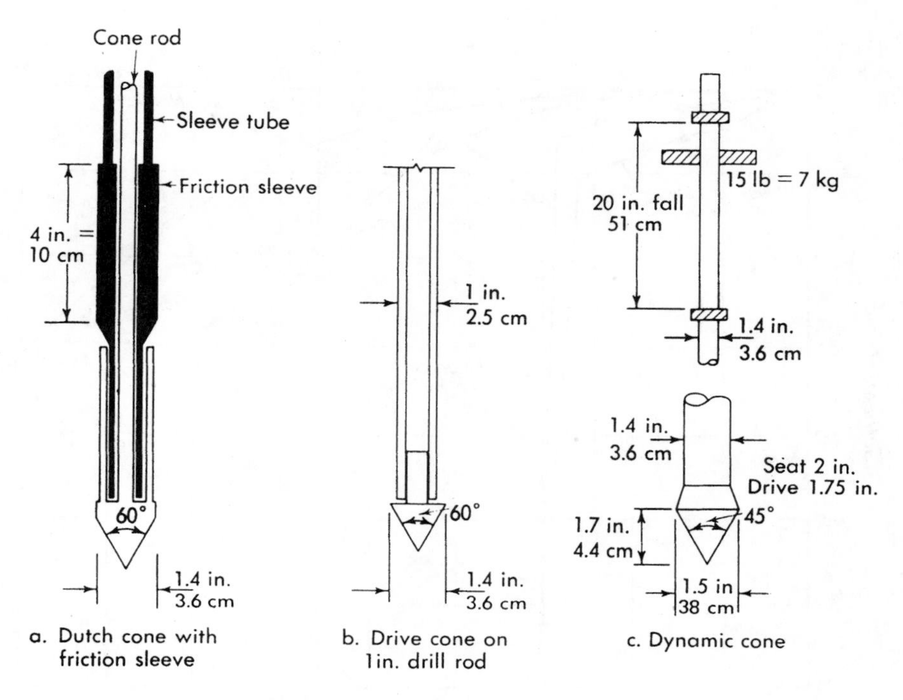

FIGURE 3.4 Cone penetrometers (after Sowers).

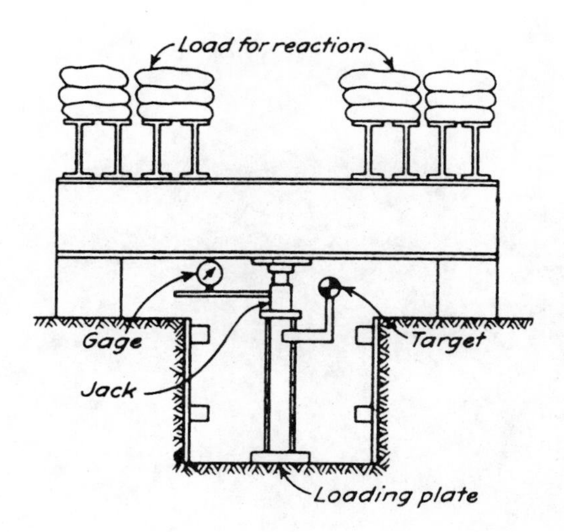

FIGURE 3.5 Arrangement for plate load test (after Peck).

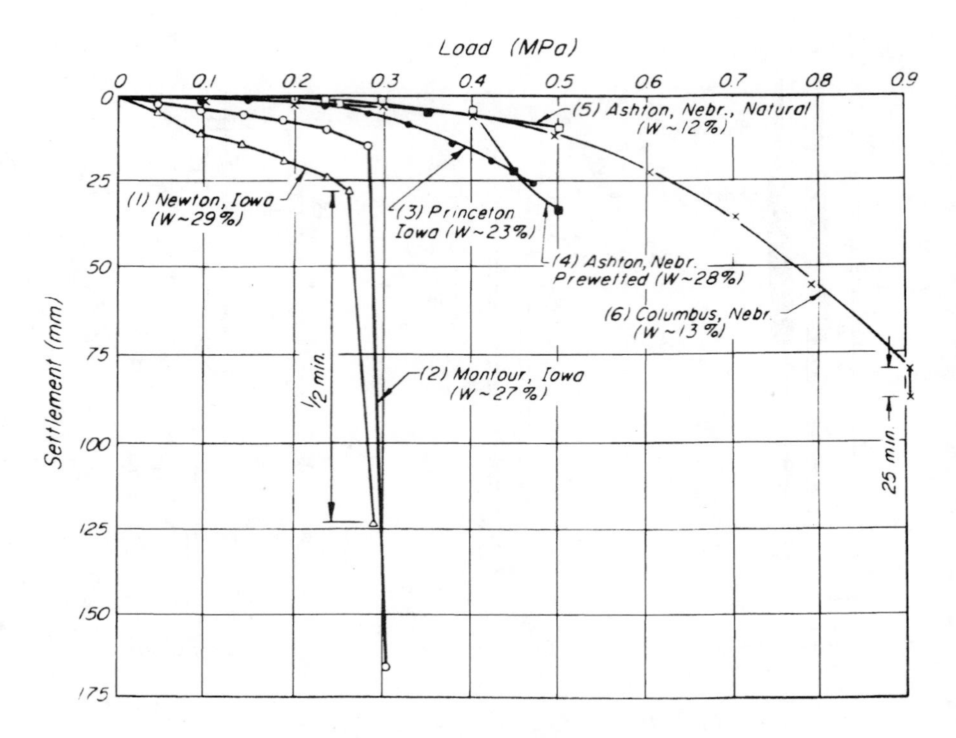

FIGURE 3.6 Results of standard load tests on loess deposit (after Peck).

FIGURE 3.7 Plate bearing test for the construction of runway.

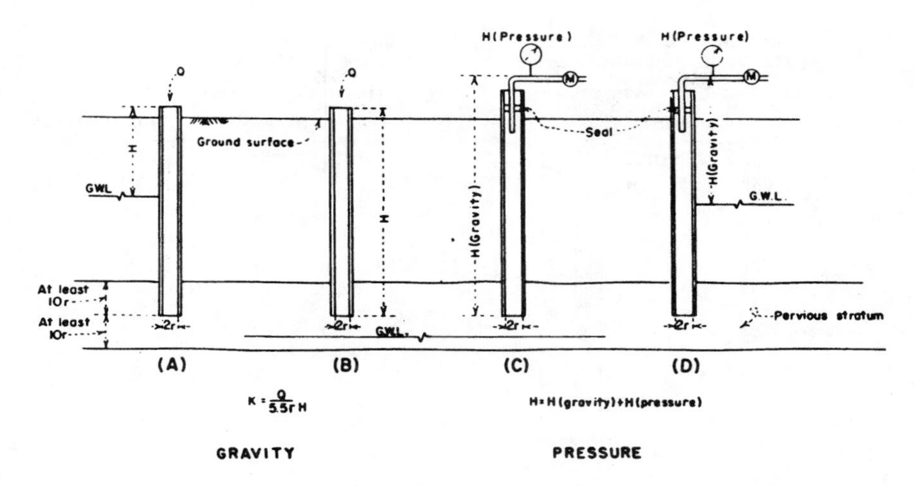

FIGURE 3.8 Open-end field permeability tests (after *Soil Manual*).

3.2.2 Packer Test

The packer test can be made both below and above the water table. It is commonly used for pressure testing of bedrock using packers, but it can be used in unconsolidated materials where a top packer is placed just inside the casing. When the packer is placed inside the casing, measures must be taken to properly seal the annular space between the casing and the drill hole to prevent water under pressure from escaping.

The permeability can be written as:

$$k = C\ Q/H \text{ (after \textit{Soil Manual})}$$

where k is in feet per year, Q is in gallons per minute, and H is the head of water in feet, acting on the test length. Measurement of the test length depends on the location of the water table as shown in Figure 3.9. Values of C for *NX* size of casing vary from 2800 for test length of 20 ft to 23,000 for test length of 1 ft.

The usual procedure is to drill the hole; remove the core barrel; set the packer; make the test; remove the packer; drill the hole deeper; set the packer again to test the newly drilled section; and repeat the test.

3.2.3 Vane Shear Test

The vane shear test is used to measure the in situ undrained shear strength of silts and clays of shallow water origin. Such tests are commonly associated with the field testing of dams and canals.

A four-bladed vane is driven into the soil at the end of a rod and the vane then rotated at a constant rate between 6 and 12°/min until the cylinder of soil contained by the blades shears. The maximum torque required for this is recorded. The blade

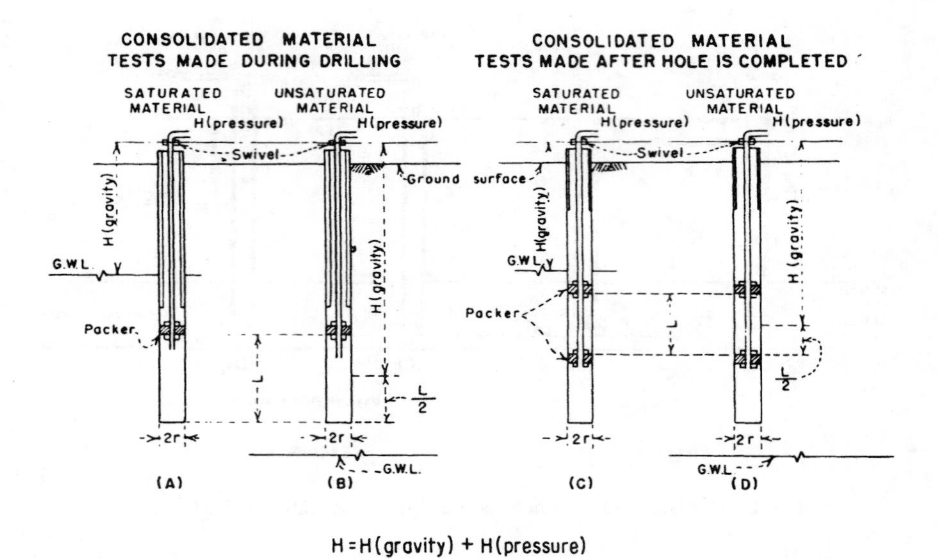

FIGURE 3.9 The packer test (after *Soil Manual*).

size varies from 75 mm wide × 150 mm long for weak soils, to 50 × 100 mm for stronger soils.

The vane rod and extension pieces are enclosed in a sleeve to prevent soil adhesion during the application of the torque. Depending on the nature of the soil, vane tests may be carried out down to depths of 100 to 250 ft (Figure 3.10).

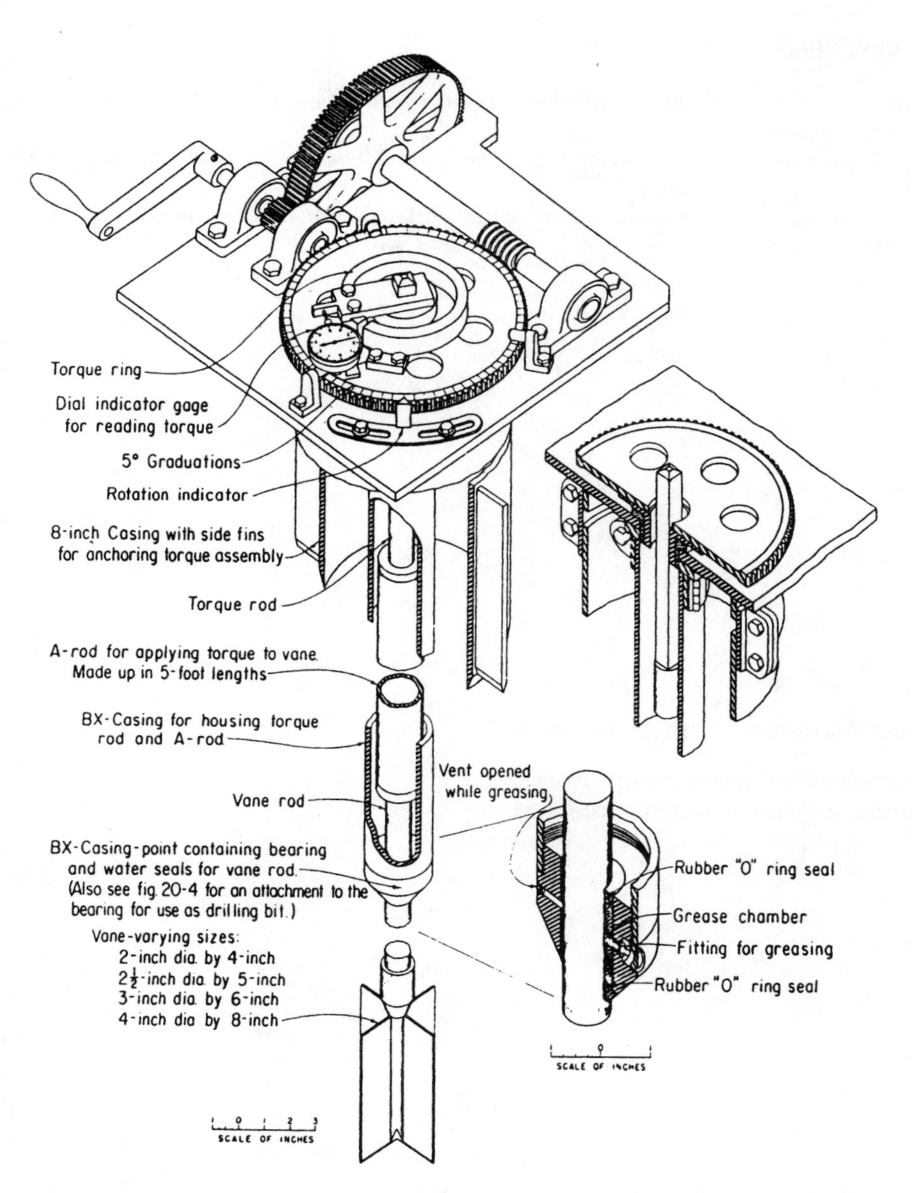
Torque ring
Dial indicator gage for reading torque
5° Graduations
Rotation indicator
8-inch Casing with side fins for anchoring torque assembly
Torque rod
A-rod for applying torque to vane. Made up in 5-foot lengths
BX-Casing for housing torque rod and A-rod
Vane rod
BX-Casing-point containing bearing and water seals for vane rod. (Also see fig. 20-4 for an attachment to the bearing for use as drilling bit.)
Vane-varying sizes:
2-inch dia. by 4-inch
2½-inch dia. by 5-inch
3-inch dia. by 6-inch
4-inch dia by 8-inch
SCALE OF INCHES
Vent opened while greasing
Rubber "O" ring seal
Grease chamber
Fitting for greasing
Rubber "O" ring seal
SCALE OF INCHES

FIGURE 3.10 Inplace vane shear test apparatus (after *Soil Manual*).

REFERENCES

R. Peck, W. Hanson, and T.H. Thornburn, *Foundation Engineering,* John Wiley & Sons, New York, 1974.

G.B. Sowers and G.S. Sowers, *Introductory Soil Mechanics and Foundations,* Collier-Macmillan, London, 1970.

U.S. Department of the Interior, Bureau of Reclamation, *Soil Manual,* Washington, D.C., 1974.

4 Classification and Identification

CONTENTS

The object of soil classification is to divide the soil into groups so that all the soils in a particular group have similar characteristics by which they may be identified.

The field engineer should be able to classify and identify the soil in the field, but field classification has to be confirmed by laboratory testing. A qualified field engineer or a technician should be able to identify the subsoil with a reasonable degree of accuracy.

It appears that both the Unified Soil Classification System and the AASHTO (American Association of State Highway and Transportation Officials) System are obvious enough for all engineers to understand, yet for a field engineer to enter the appropriate symbols in the drill log can sometimes be difficult. A field engineer has to examine the soil as it comes out from the auger, locate the next bore hole, measure the water, and sometimes help the driller; he has little time left to study the accuracy of the log. It is up to the engineer in charge to discuss the project with the field engineer, analyze the findings, and make the necessary changes before issuing the final report to the client.

4.1 FIELD IDENTIFICATION AND CLASSIFICATION

The earliest soil classification system is the "Texture System." This system is based on soil grain size alone. It is still used in the agricultural field, though geotechnical engineers have long abandoned the system. The FAA system initiated by the Federal Aviation Administration was the first to be developed by the engineers. The only systems in use by the geotechnical engineers at present are the AASHTO System and the Unified Soil Classification system. The latter has gradually replaced all others.

4.1.1 Unified Soil Classification System

The system was adopted in 1952 by the Bureau of Reclamation and the Corps of Engineers, with Professor A. Casagrande as consultant. Soils are categorized in groups, each of which has distinct engineering properties. The criteria for classification are

1. Percentage of gravel, sand, and fines in accordance with grain size
2. Shape of grain size curve for coarse grain soils
3. Plasticity and compressibility characteristics for fine grained soils and organic soils. The letter symbols are as follows:

G	Gravel and gravelly soils
S	Sand and sandy soils
M	Silt
C	Clay
O	Organic soils
Pt	Peat
W	Well graded
P	Poorly graded
H	High Plasticity
L	Low Plasticity

In borderline cases, a combination of symbols is used. The Unified Soil Classification System is a milestone in soil mechanics.

For the use of the field engineer, the Bureau of Reclamation and the Corps of Engineers have published "Field Identification Procedures" and guidelines so that all field logs can be interpreted using the same language. The system was revised in 1952, and is presented in Tables 4.1a, 4.1b, and 4.2.

TABLE 4.1a
Unified Classification System — Group Symbols for Gravelly Soils

Group Symbol	Criteria
GW	Less than 5% passing No. 200 sieve; $C_u = D_{60}/D_{10}$ greater than or equal to 4; $C_c = (D_{30})^2/(D_{10} \times D_{60})$ between 1 and 3
GP	Less than 5% passing No. 200 sieve; not meeting both criteria for GW
GW-GM	Percent passing No. 200 sieve is 5 to 12; meets the criteria for GW and GM
GW-GC	Percent passing No 200 sieve is 5 to 12; meets the criteria for GW and GC
GP-GM	Percent passing No 200 sieve is 5 to 12; meets the criteria for GP and GM
GP-GC	Percent passing No. 200 sieve is 5 to 12; meets the criteria for GP and GC

TABLE 4.1b
Unified Classification System — Group Symbols for Sandy Soils

Group Symbol	Criteria
SW	Less than 5% passing No. 200 sieve; $C_u = D_{60/}D_{10}$ greater than or equal to 6; $C_c = (D_{30})^2/(D_{10} \times D_{60})$ between 1 and 3
SP	Less than 5% passing No. 200 sieve; not meeting both criteria for SW
SM	More than 12% passing No. 200 sieve; Atterberg's limits plot below A-line or plasticity index less than 4 (Figure 4.1)
SC	More than 12% passing No. 200 sieve; Atterberg's limits plot above A-line; plasticity index greater than 7 (Figure 4.1)
SC-SM	More than 12% passing No. 200 sieve; Atterberg's limits fall in hatched area marked CL-ML (Figure 4.1)
SW-SM	Percent passing No. 200 sieve is 5 to 12; meets criteria for SW and SM
SW-SC	Percent passing No. 200 sieve is 5 to 12; meets criteria for SW and SC
SP-SM	Percent passing No. 200 sieve is 5 to 12; meets criteria for SP and SM
SP-SC	Percent passing No. 200 sieve is 5 to 12; meet the criteria for SP and SC

(after Das)

TABLE 4.2
Unified Classification System — Group Symbols for Silty and Clayey Soils

Group Symbol	Criteria
CL	Inorganic: LL < 50; PL > 7; plots on or above A-line (Figure 4.1)
ML	Inorganic: LL < 50; PL < 4; or plots below A-line (Figure 4.1)
OL	Organic: (LL — oven dried)/(LL — not dried) < 0.75; LL < 50
CH	Inorganic: LL more than 50; PL plots on or above A-line
MH	Inorganic; LL more than 50; PL plots below A-line (Figure 4.1)
OH	Organic; (LL — oven dried)/(LL — not dried) < 0.75; LL > 50
CL-ML	Inorganic; plot in the hatched zone (Figure 4.1)
Pt	Peat, muck, and other highly organic soils

(after Das)

For proper classification according to the above tables, the following should be noted:

1. Percent of gravel — that is, the fraction passing the 3-in. sieve and retained on the No. 4 sieve
2. Percent of sand — that is, the fraction passing No. 4 sieve and retained on the No. 200 sieve
3. Percent of silt and clay — that is, the fraction finer than the No. 200 sieve

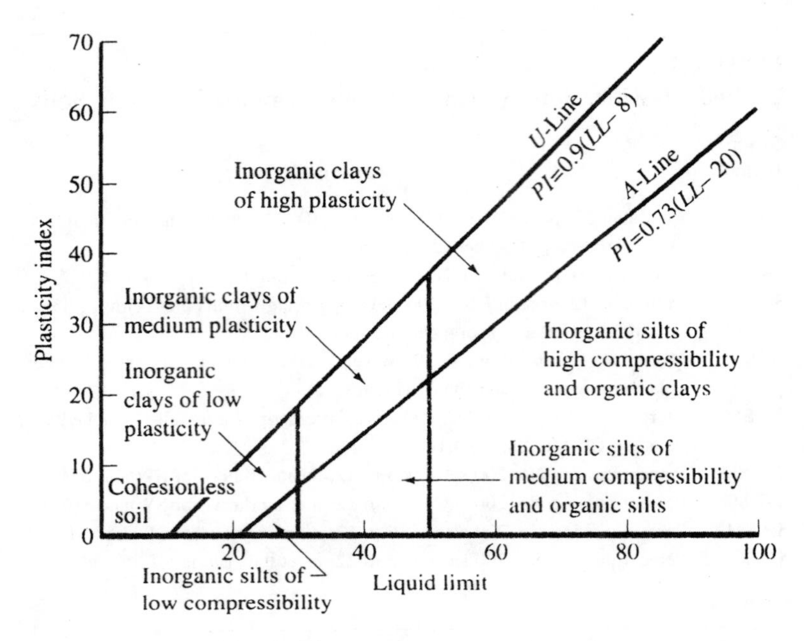

FIGURE 4.1 Plasticity Chart (after Das).

4. Uniformity coefficient (C_u) and the coefficient of curvature (C_c)
5. Liquid limit and plasticity index of the portion of soil passing the No. 40 sieve

The most commonly encountered granular soils are GC-GM, GP-GC, SP, SC, SP-SC. For cohesive soils the CL,CH, and the CL-CH are most commonly observed by consultants.

4.1.2 Highway Soil Classification System

About 1928, the Bureau of Public Roads introduced a soil classification system still widely used by highway engineers. All soils were divided into eight groups designated by the symbols A-1 through A-8. The soil best suited for the subgrade of a highway is designated as A-1. The material is largely a well-graded sand and gravel, but containing a small amount of binder. All other soils were grouped roughly in decreasing order of stability. This system is shown in Table 4.3.

In the AASHTO System, the inorganic soils are classified in seven groups corresponding A-1 through A-7. These in turn are divided into a total of 12 subgroups. Organic soils are classified as A-8. Any soil containing fine-grained material is further identified by its group index; the higher the index, the less suitable the soil. The group index is calculated from the formula

$$GI = (F-35)[(0.2+0.005)(LL-40)] + 0.01(F-15)(PI-10)$$

where F = percentage passing No. 200 sieve expressed as a whole number.

The system is complicated; consultants prefer to use the Unified System. However, the consultants should be familiar with the AASHTO System when the projects involve the federal or the state government.

4.2 IDENTIFICATION AND DESCRIPTION

Accurate and precise description of the subsoil conditions in the field by the field engineers is not easy. A freshman field engineer tends to over-log the subsoil strata revealed in the bore hole. As the soil comes out of the auger, there are some differences in each foot of the material. Consequently, the inexperienced engineer tends to log the hole with such frequent changes of stratum that the draftsman has a difficult time plotting the data. At the same time, drill logs without any descriptions are often seen.

Logging of test holes is an art. It takes a great deal of practice and experience before the field engineer or a technician is able to master such a skill. Bearing in mind that subsoils are never uniform, the field engineer logs the holes for engineering purposes, not for research. Inconsistency is to be expected. However, if the field log fell into the hands of an attorney, the engineer can at once smell trouble. It is sometimes better to destroy the field log after completion of the project.

4.2.1 WATER MEASUREMENT

Moisture content of soils as they are taken from the ground should be recorded. They are generally designated as dry, moist, or wet. These terms are relative depending on the type of soil. For fairly clean sands, a moisture content of about 12% is considered as wet, while for clay this moisture content is considered to be moist.

It is essential that the field engineer record the water level found in the bore holes. In the case of sandy soils, when the augers are removed rapidly from the bore hole, an upward hydraulic gradient is created in the sand beneath the drill hole. Consequently, the sand may become "quick" and the relative density may be greatly reduced. The penetration resistance value will accordingly be much lower than that corresponding to the relative density of the undisturbed sand. Care is required to ensure that the water level in the drill hole is always maintained at or slightly above that corresponding to the piezometric level at the bottom of the hole.

The importance of water in geotechnical engineering will be discussed further in later chapters.

4.2.2 LABORATORY CLASSIFICATION

Although most classification of soils can be performed in the field, the Unified Soil Classification System has provided for precise delineation of the soil group by gradation analyses and Atterberg limits tests in the laboratory.

Within the sand and gravel groupings, soils containing less than 5% finer than the No. 200 sieve size are considered "clean." A clean sand having both a coefficient of uniformity greater than 4 and a coefficient of curvature between 1 and 3 is in the SW group; otherwise, it is a poorly graded sand SP.

TABLE 4.3
Classification of Highway Subgrade Material

General classification	**Granular materials (35% or less of total sample passing No. 200)**						
Group classification	**A-1**			**A-2**			
	A-1-a	**A-1-b**	**A-3**	**A-2-4**	**A-2-5**	**A-2-6**	**A-2-7**
Sieve analysis (percent passing)							
No. 10	50 max.						
No. 40	30 max.	50 max.	51 min.				
No. 200	15 max.	25 max.	10 max.	35 max.	35 max.	35 max.	35 max.
Characteristics of fraction passing No. 40							
Liquid limit				40 max.	41 min.	40 max.	41 min.
Plasticity index	6 max.		NP	10 max.	10 max.	11 min.	11 min.
Usual types of significant constituent materials	Stone fragments, gravel, and sand		Fine sand	Silty or clayey gravel and sand			
General subgrade rating	Excellent to good						

General classification	**Silt-clay materials (more than 35% of total sample passing No. 200)**			
Group classification	*A-4*	*A-5*	*A-6*	*A-7* *A-7-5*[a] *A-7-6*[b]
Sieve analysis (percent passing)				
No. 10				
No. 40				
No. 200	36 min.	36 min.	36 min.	36 min.
Characteristics of fraction passing No. 40				
Liquid limit	40 max.	41 min.	40 max.	41 min.
Plasticity index	10 max.	10 max.	11 min.	11 min.
Usual types of significant constituent materials	Silty soils		Clayey soils	
General subgrade rating	Fair to poor			

[a] For A-7-5, $PI \leq LL - 30$
[b] For A-7-6, $PI > LL - 30$

(after Das)

"Dirty" gravels or sands are those containing more than 12% fines, and they are classified as either clayey or silty by results of their Atterberg limits tests as shown in the plasticity chart.

If a soil is determined to be fine grained by using the grain size curve, it is further classified into one of the six groups by plotting the results of Atterberg limits tests on the plasticity chart. Fine-grained soils with PI greater than 7 and above the "A" line (Figure 4.1) are CL or CH, depending on whether their liquid limits are less than 50% or more than 50%, respectively.

Similarly, fine-grained soils with PI less than 4 or below the "A" line are ML or MH, depending on whether their liquid limits are less than or more than 50%, respectively. All geotechnical engineers are familiar with the plasticity chart and the significance of the "A" line as shown in Figure 4.1.

The types of soil with which most geotechnical engineers are concerned are not as complicated as presented in the chart. The following are observed:

1. For the coarse grain soil, it must be emphasized that in order to justify the symbol SP or SW, the percentage of fines should be less than 5%; otherwise, dual symbols should be used. Also, to justify the symbol SW or GW, the gradation must fulfill the uniformity coefficients and coefficient of curvature requirements. Such soils are seldom encountered. Most, if not all, foundation soils encountered in foundation design are of GP or SP group.
2. For the fine-grained soils, those soils below the "A" line such as OH, MH, and OL are seldom encountered. Peat soils, of course, are quite rare. Engineers should direct their attention to the most commonly found soils such as CL, ML, and CH. The differentiation between CL and ML is of special importance.
3. Dual symbols should be used with care and only when necessary, especially in the course of visual classification. The use of CL-SC or CL-ML is acceptable, but the use of CL-SP or CH-SM is inexcusable. Also, no attempt should be made to use three symbols. The use of dual systems indicates the engineer is doubtful of his classification, and the use of triple symbols indicates that he has no idea what he is doing.
4. For stiff clays, it is possible to predict their behavior by visual examination. Very stiff clays with high plasticity generally exhibit swelling potential while the low density, calcareous clays with porous structure indicate high settlement or collapsing characteristics.

4.2.3 BEDROCK

The description of bedrock in geotechnical engineering is a little different from the terms used by the geologist. However, the underlying meanings are the same. Shale bedrock in the Rocky Mountain area is familiar to most consulting engineers. Besides consistency and moisture content, the degree of cementation, the angle of dip, and the stratification should be recorded.

TABLE 4.4
Pentetration Resistance and Consistency of Shale

Penetration Resistance (blows/ft)	Description
Less than 20	Clay (weathered claystone}
20–30	Firm
30–50	Medium hard
50–80	Hard
over 80	Very hard

(after Peck, Hanson, and Thornburn)

A guideline for describing the consistency of shale bedrock with respect to penetration resistance is shown in Table 4.4.

In the east side of the Front Range area bedrock formations have been turned up by the uplifting. This unique bedrock formation is noteworthy. The bedding planes of the bedrock, instead of lying flat, dip steeply at an angle greater than 30° from horizontal. The Colorado Geological Survey designated the area as the "Dipping Bedrock Overlay District." The geological cross-section is shown on Figure 4.2. The effects of such formation are that the rock layers with distinctly different swelling characteristics are found in close proximity to each other along or near the ground surface in the form of thin zones. These zones respond differently to the increase in moisture, with the more expansive claystone undergoing more swelling and vertical uplift than the adjacent beds. The resulting effect is differential, often destructive heave. A second effect of upturned bedding is the development of higher-than-normal lateral swelling pressures. When exposed at or near the ground surface, claystone bedrock is free of its burial confining pressures and will try to attain its original thickness by pulling water into its structure and expanding.

Recently, the dipping claystone shale stratification in the Denver Front Range area has received much attention as it had great effect on the performance of the lightly loaded structures. Dipping claystone bedrock formation allows surface water to enter the lower highly swelling soil. It also creates a very complicated and erratic subsoil condition that may require a different approach in design.

Differentiation between claystone and sandstone should be easy, while claystone and siltstone are sometimes difficult to identify. Color may not be important in engineering, but it does offer assistance for identification. Siltstone in the Laramie, Wyoming area is generally red. Denver blue shale derived its name from its dark gray color, while chocolate brown claystone generally possesses high swelling potential.

Interlayered shale bedrock consists of alternative layers of claystone. The recognition of such strata is important. Sometimes, the field engineer logs bedrock as non-expansive sandstone and neglects to record the thin layer of highly expansive claystone that turns out to be the key material in the foundation design.

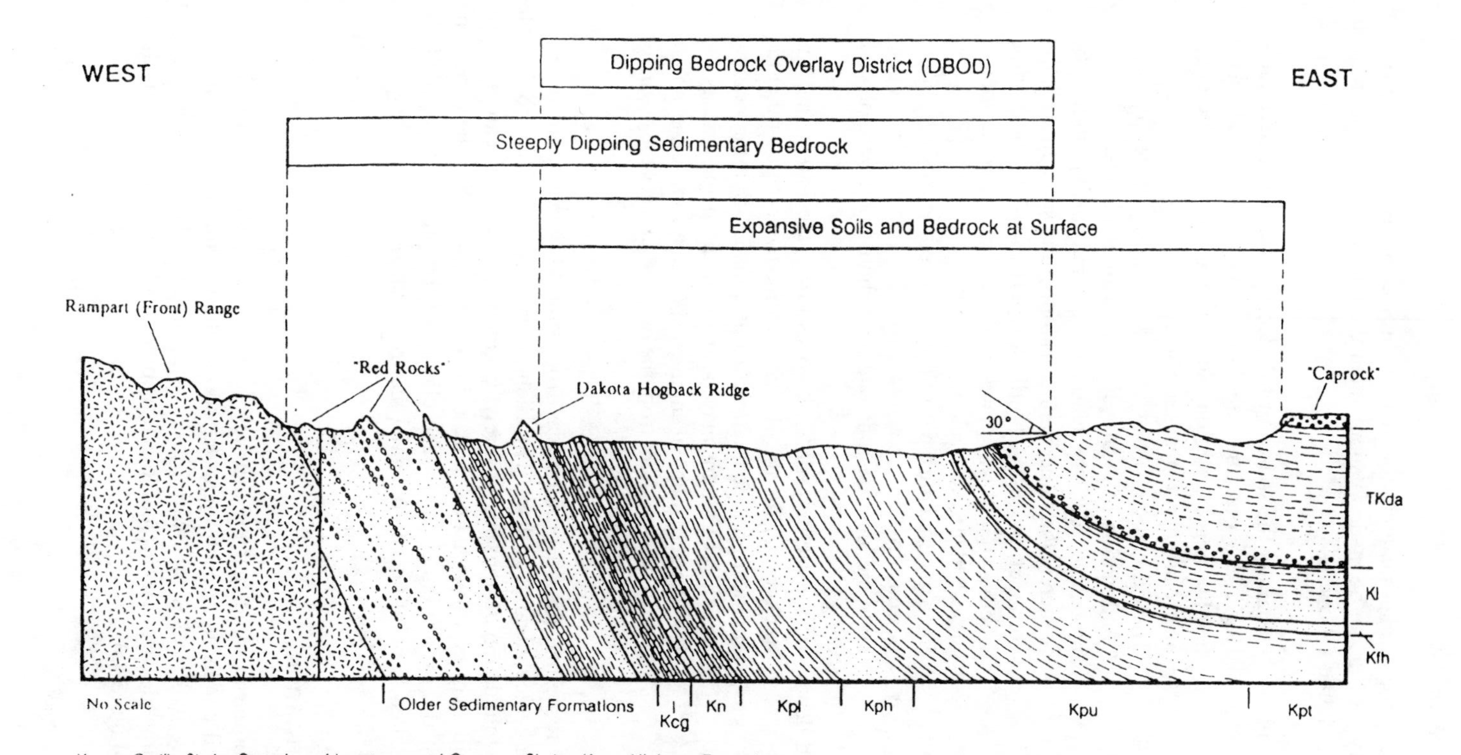

FIGURE 4.2 Geological cross-section showing dipping bedrock formation along the Denver Front Range area (after CGS).

4.2.4 Man-made Fill

Complete description of man-made fill found at the project site is imperative as very little laboratory testing can be made to supplement the findings. Description of the fill should include the following:

Source of fill	Generally the source of fill cannot always be determined. However, sometimes the source is obvious, such as a neighboring excavation. Some probing by the field engineer can often reveal the source.
Type of fill	The fill can consist of clean soils, trashy materials, or a combination of clean soil and trashy materials.
Clean fill	For clean fills, the classification and testing should be treated in the same manner as for natural soils. In addition, compaction tests should be performed to determine whether the fill was placed under controlled conditions or merely placed at random.
Organic fill	Such material must be entirely removed. The owner should be consulted on the disposal site. Organic matter can generate methane gas; the engineer should be aware of such danger.
Extent and depth	It is extremely difficult to estimate the extent and the thickness of the fill. However, some owners insist that the engineer provide them with a percentage. Bearing in mind that the consultant may be liable if the percentage is too low, the geotechnical engineer should in some cases obtain a written agreement from the owner that frees the engineer from any liability.

Failure to recognize the fill problem has from time to time presented serious trouble for the geotechnical consultant. The client, being a layman, may not be able to understand foundation design, but he surely can recognize fill. The extent, depth, and type of fill may be difficult to delineate in the test holes, but these become apparent when the excavation is open. The client will be greatly annoyed if the engineer has failed to point out the potential hazard or the added construction cost due to the presence of fill.

4.2.5 Presentation

It should be borne in mind that the drill log is supposed to record the general or average subsoil condition in each major subsoil stratum. Using a description such as "sandy clay with lenses of clean gravel" will simplify the drill log and still maintain accurate description. Typical logs of exploratory holes used by consultants are shown in Figure 4.3.

LOG OF HOLE 2-3 JOB NAME Meadows Ped. Pass No. 2 DATE/TIME 7-6-8

JOB NO. 140586 HOLE LOCATION NE Bridge Abutment ENGINEER DI

HOLE ELEVATION 100.5 DRILLING METHOD 4" CFA RIG/DRILLER #17/Tony

DEPTH FROM	DEPTH TO	UNIFIED SYMBOL	SOIL DESCRIPTION AND DRILLING CONDITIONS
0	2"		Topsoil
2"	1'	FILL	Clay, fine to med grained, sl. silty, trace gravel, med. pl[illegible] med. stiff, moist, d. brown
1'	16'	FILL	Clay, very sandy, silty to primarily sl. silty, low to primarily med. plastic, fine to coarse grained, occ. sl. gravelly, med to stiff, mix brown, primarily lt. moist to moist
			Note: Hit sprinkler line @ 1'

DEPTH TO TOP (FEET)	BLOW COUNT	TYPE SAMPLE	SOIL SYM.	PLASTIC LOW	PLASTIC MEDIUM	PLASTIC HIGH	% FINES (− 200)	GRADING (+ 200) FINE	MEDIUM	COARSE	GRAVEL	CONSISTENCY DENSITY HARDNESS COMPACTNESS	PEN. (TSF)	MOISTURE DRY	MOIST	WET	COLOR	POROUS	CALC.	RECOVERY	OTHER (CEMENTED, FRACTURED, STRATIFIED, BLOCKY, QUALITY OF SAMPLE, ETC.)
1	7/12	CAL	FILL	X			70	X	X	X	sl	Med.			X		Mix brn			X	V. Sandy, s. silty
2	8/12	↓	FILL		X		70	X	X	X	sl	Stiff			lt		Mix brn			X	V. Sandy, sl. s[illegible]
					X																
4	8/12	CAL	FILL		X		70	X	X	X	sl	Stiff			lt		Mix brn			X	V. Sandy, sl. sil[illegible]
5	7/12	↓	FILL		X		65	X	X	X	sl	Med			lt		Mix brn			X	V. Sandy, sl. silt[illegible]
7	6/12	CAL	FILL		X		70	X	X	X		Med.			lt		Mix brn			X	V. Sandy, sl. s[illegible]
8	10/12	↓	FILL		X		70	X	X	X		Stiff			lt		Mix brn			X	V. Sandy, sl. s[illegible]
10	6/12	CAL	FILL		X		70	X	X	X		Med.			lt		Mix brn			X	V. Sandy, sl. silt[illegible]
11	10/12	↓	FILL		X		65	X	X	X	sl	Stiff			lt		Mix brn			X	V. Sandy, sl. si[illegible]

CAL = CALIFORNIA
SS = SPLIT SPOON
ST = SHELBY TUBE
LD = LARGE DISTURBED (>2 LB.)
SD = SMALL DISTURBED (<2 LB.)
P = PITCHER
HD = HAND DRIVE

WATER LEVEL	dry		
DATE	7-6		
TIME	DRILL		

DEPTH TO BEDROCK —
DEPTH TO CAVE —
PVC CASING (DEPTH/DIAM.) —
BOTTOM HOLE DEPTH 16

FIGURE 4.3 Typical logs of exploratory holes used by consultants.

After the completion of soil classification, the next step is the plotting of the log. This is usually left in the hands of a draftsman or done by using a computer. Close checking of the log should be made by the project engineer. The following should be included:

Unified soil legend — By using the soil legend suggested by the Bureau of Reclamation, a soil engineer or an architect is able to identify the soil without reading the description. The types of soil presented in this legend are limited to only eight, with three additional symbols for fill and six for bedrock. Complicated and detailed classifications are not considered necessary in general exploration and sometimes may confuse the issue. Typical logs are shown in Figure 4.3.

Plotting — All test holes should be plotted according to elevation. When elevations are not taken, notes and explanations should be given. A horizontal line should be drawn across the log, indicating the proposed floor level. In this manner, a concise idea on the subsoil conditions immediately beneath the footings can be obtained. Without the proposed floor level, it will be necessary to assume one or several possible floor levels and build the recommendations around the assumptions. Typical soil legends and symbols are shown in Figure 4.4.

Water level — The water table is an integral part of a soil log. The depth of the water table should be carefully recorded. Stabilized water table conditions can generally be obtained in the test hole after 24 hours. Such records should be plotted. In cohesive soils due to their low permeability, no water or low water table conditions are generally recorded. The field engineer should record the water level with clear explanations.

Others — The log should also include such data as the date of drilling, the location of bench mark, type of drilling equipment, climate condition, and the engineer's name.

TOPSOIL, OR ORGANIC SOIL.

CLAY, CL CL–CH, CH, FIRM.

SILT, ML ML–CL FIRM.

SAND, CLEAN, SP, SW, SP, PERVIOUS.

SAND, CLAYEY, SC, SC–SM, IMPERVIOUS.

SAND, SILTY, SM, AM–SC, IMPERVIOUS.

GRAVEL, CLEAN, WITH OR WITHOUT SAND, GW–SW, GP–SP, PERVIOUS.

GRAVEL, CLAY WITH SAND, GC–SC.

GRAVEL, SILTY WITH SAND, GM–SM

LESS COMMON

ALTERNATING STRIPS INDICATES LAYERED OR VARING CL–SC.

CLAY OR SILT SYMBOL SLANTING IN OPPOSITE DIRECTION OR STRAIGHT LINE INDICATES VERY LOOSE.

INDICATES BOULDERS, LARGE COBBLE.

SYMBOLS

GRADUAL CHANGE IN MATERIAL

10/12 INDICATES THAT 13 BLOWS OF A 140–POUND HAMMER FALLING 30 INCHES WERE REQUIRED TO DRIVE THE SAMPLER 12 INCHES.

UNDISTURBED SHELBY TUBE SAMPLE

DISTURBED STANDARD SPLIT SPOON

DEPTH AT WHICH HOLE CAVED

INDICATES DEPTH TO FREE WATER AND

5 INDICATES DEPTH TO FREE WATER AND NUMBER OF DAYS AFTER DRILLING THAT MEASUREMENT WAS TAKEN.

CONCRETE OR ASPHALT PAVING AND BASE COURSE.

FILL MAN–MADE, LOOSE OR UNKNOWN

FILL SPECIAL SYMBOL, DENSE, CONTROLLED, SLAG, ETC.

BEDROCK

WEATHERED CLAYSTONE, CL–CH, UNDER 20 BLOWS

WEATHERED SANDSTONE, SC–SP–SM–SW, UNDER 20 BLOWS

CLAYSTONE BEDROCK, FIRM MEDIUM HARD 20–50 BLOWS

SANDSTONE BEDROCK, FIRM MEDIUM HARD 20–50 BLOWS

CLAYSTONE–SANDSTONE, OR SILTSTONE BEDROCK HARD TO VERY HARD OVER 50 BLOWS.

GRANITE, SCHIST, GNAISS OR SIMILAR HARD COMPETENT ROCK

NOTES

1. TEST HOLES WERE DRILLED __________ WITH A 4–INCH DIAMETER CONTINUOUS FLIGHT POWER AUGER.
2. ELEVATIONS ARE APPROXIMATE HA D LEVEL REFER TO B.M. SHOWN ON FIG. 1
3. NO FREE WATER WAS FOUND IN TEST HOLES AT THE TIME OF DRILLING
4. WC = WATER CONTENT (%)
   DD = DRY DENSITY (pcf)
   UC = UNCONFINED COMPRESSIVE STRENGTH (psf)
   LL = LIQUID LIMIT (%)
   PI = PLASTICITY INDEX (%)

GENERAL			
NON–COHESIVE SANDS		COHESIVE CLAY	
1–10	LOOSE	0–2	VERY SOFT
10–30	MEDIUM DENSE	2–4	SOFT
30–50	DENSE	4–8	MEDIUM SOFT
50+	VERY DENSE	8–16	STIFF
		8–16	VERY STIFF

FIGURE 4.4 Typical soil legend and symbols used by consultants.

REFERENCES

Arthur Casagrande. Classification and Identification of Soils, Trans. ASCE 113, New York, 1948.

Corps of Engineers, Department of the Army, VII. I.

B.M. Das, *Principles of Geotechnical Engineering,* PWS Publishing, Boston, 1994.

R. Peck, W. Hanson, and T.H. Thornburn, *Foundation Engineering,* John Wiley & Sons, New York, 1974.

U.S. Department of the Interior, Bureau of Reclamation, *Soil Manual,* Washington, D.C., 1974.

5 Laboratory Soil Tests

CONTENTS

Soil testing is essential in establishing the design criteria. Distinction should be made between the needs of the consulting engineer and those of the research engineer. For a practicing engineer, the purpose of laboratory testing is mainly to confirm his or her preconceived concept. Exotic laboratory equipment and refined analyses are in the realm of the research engineer or the academician. Neither time nor budget will allow the practicing engineer or the consultant to follow the researcher's procedures.

An experienced consulting geotechnical engineer usually has an idea as to the type of foundation and the design value for the assigned project before the commencement of laboratory testing. Such a concept is usually derived from the field drilling log, field penetration data, visual examination of the sample, and the experience of the area.

To most geotechnical engineers, the difference between sand and clay is apparent. However, in the case of sand, the symbols SW should be used with care, since clean sands as the symbol implies are rarely encountered. In the case of fine-grained soils, the difference between "clay" and "silt" is not apparent visually; a plasticity test will be required.

The crude and the most elementary method used by the engineer to identify soil is to take a small lump of soil and roll it on the palm after spitting on it. If color appears on the palm, it is likely to be CL or CH. Otherwise, it is probably silt. For granular soils, one can chew the soil between the teeth. A gritty feeling indicates sandy soil, probably SC.

5.1 SCOPE OF TESTING

The extent of soil testing required for a project varies. It depends on the type of client; the importance of the project; the funding available; the time required; and to some extent, the capability of the consultant's laboratory.

5.1.1 Standard Tests

The commonly conducted soil tests by the consulting engineering firms consist of the following:

- Moisture content and index tests
  - Moisture content
  - Liquid limit
  - Plasticity limit
  - Shrinkage limit
- Density and specific gravity
- Particle size analysis
- Compaction test
- Shear strength
  - Triaxial shear test
  - Direct shear test
  - Unconfined compression test
- Compressibility and settlement
  - Consolidation test
  - Swell test
- Permeability test (Figure 5.1)

All the above tests are well described in almost all soil mechanics literature. In addition, most of the test procedures are now listed in ASTM as standard. Improvements are necessary in many areas, especially in the subject of swelling soils.

5.1.2 Minimum Testing Capability

A consulting soil engineer usually starts with minimum financial backing and cannot afford to buy all the elaborate testing apparatus found in the specialized catalogs. Some of the successful consulting firms want to expand their operations to another location but hesitate because they are unable to purchase the necessary costly testing apparatus. In fact, in the U.S., only government organizations such as the Bureau of Reclamation or the Corps of Engineers can afford to purchase all the up-to-date new items.

Visitation to several soil laboratories in Asia and in the Middle East found them modern, well equipped, and unusually clean. Clean apparatus indicated that the facilities were seldom used. Other governmental institutes located short distances from each other had duplicate equipment. Sharing the use of high-cost equipment was never considered.

FIGURE 5.1 Permeability test.

For starting geotechnical engineers, the following minimum apparatus is recommended:

A drying oven (home baking oven can also be used)
A set of sieves (nothing wrong with hand shaking instead of a mechanical shaker)
One unconfined compression test apparatus (hand operated)
Four simplified consolidation apparatuses (locally made)
A set of graduated glass cylinders
A set of Proctor cylinder and hammer
Large and small scale balance

The above apparatus can be obtained at minimum cost; additional items can be purchased as business grows. Consolidation or swell testing equipment is necessary for providing the key data for establishing the foundation design criteria. A train of consolidation apparatus is sometimes necessary to shorten the time of testing. The simplified consolidation apparatus is shown in Figure 5.2.

FIGURE 5.1 (continued)

5.2 INTERPRETATION OF TEST RESULTS

Laboratory testing of disturbed and undisturbed soil samples can be performed in most soil testing laboratories by trained laboratory technicians. Laboratory test results are reliable only to the extent of the condition of the sample. Results of testing on badly disturbed samples or samples not representative of the strata are not only useless, but also add confusion to the complete program. In some geotechnical reports, the writer may include everything the laboratory technician puts in front of him for the sole purpose of increasing the volume of the report. This practice is especially common in Asian countries where soil reports are not critically reviewed.

A reasonably good sample can be obtained when driving into shale bedrock or stiff clays. Auger drilling in most cases can be successfully conducted in such soils. For bedrock such as limestone and granite, rotary drilling is necessary and rock cores can be obtained. Core samples are brought up by the drill and can be visually examined. The general characteristics, in particular the percentages of recovery, are of importance to the foundation design and construction cost.

Equally important to testing of representative samples is the frequency of testing. Testing of a few samples in a single project and basing the final analysis on such

FIGURE 5.2 Modified Consolidation Test.

testing is not only undesirable, but also dangerous. By testing only a few samples, the swelling or collapsing characteristics may be missed and erroneous conclusions drawn. Too little testing is sometimes worse than no testing at all.

An experienced geotechnical consultant should be able to screen the laboratory test results and exclude the dubious ones, the unreasonable ones, and the defective tests. After such screening, the consultant is justified in using the data to determine the maximum and the minimum value. From such values, the average value used in design can be established.

To fulfill the above procedure, it is obvious that a number of samples taken from many test borings is required. Bear in mind that the art of soil mechanics is based on the use of average value instead of the highest or the lowest value. Judgment always comes before numerical figures.

5.2.1 Swell Test

The most important laboratory test on expansive soils is the swell test. The standard one-dimensional consolidation test apparatus can be used. A standard consolidometer can accommodate a remolded or undisturbed sample from 2 to 4.25 in. diameter and from 0.75 to 1.25 in. thickness. Porous stones are provided at each end of the specimen for drainage or saturation. The assembly is placed on the platform scale

FIGURE 5.2 (continued)

table and the load is applied by a yoke actuated by a screw jack. The load imposed on the sample is measured by the scale beam, and a dial gage is provided to measure the vertical movement.

The advantage of such arrangement is that it is possible to hold the upper loading bar at a constant volume and allow the measurement of the maximum uplift pressure of the soil without a volume change. This requires a constant load adjustment by an operator. An advanced scheme is an automatic load increment device that measures swelling pressure without allowing volume change to take place.

The consolidometer can also be used to measure the amount of expansion under various loading conditions. Since swelling pressure can be evaluated by loading the swelled sample to its original volume, it is simple to convert the platform-scale consolidometer into a single-lever consolidation apparatus. Such a modified consolidometer can be made locally at low cost. The average soil laboratory should have a train of such apparatuses to speed up the testing procedure.

It is important for the geotechnical engineer not to confuse "swell" with "rebound." All clays will rebound upon load removal, but not all clays possess swelling potential. The use of graduated cylinders to measure the swelling potential of clay upon saturation is not a standard test. Such a test has been abandoned and should not be repeated.

FIGURE 5.3 Consolidation Test.

5.2.2 Consolidation Test

The type of soil test that has received the most attention is the consolidation test (Figure 5.3). Ever since Terzaghi advanced the theory of consolidation, the procedure of the consolidation test has been improved and corrected many times. Research includes the size of sample, the loading condition, the speed of loading, the duration of loading, and the drainage condition. With the aid of computer control, a consolidation setup can become the showpiece in the consultant's laboratory. One lists the error-prone areas in the sampling and testing procedures as follows:

Ring friction — The effective stress actually applied to the soil is reduced due to friction between the consolidation ring and the sides of the soil specimen.

Flow impedance — The porous stones above and below the specimen must be sufficiently fine grained to prevent clogging by the soil particles.

Sample disturbance — Sample disturbance is the result of a combination of a number of factors: the sampler effect, transport and storage effect, and sample preparation.

Rapid loading — With a daily reloading cycle, the measured values of compressibility are higher than when using either hourly or weekly cycles.

Consulting engineers realize that the use of consolidation test results for the estimation of foundation settlement is by no means accurate. Among the many inconsistencies, the most undetermined factor is that all consolidation tests are under one-dimensional conditions, whereas the actual site conditions are not. Under one-dimensional conditions, the lateral strain is zero, and the initial increase in pore pressure is equal to the increase in total stress.

Another major difference between the laboratory consolidation results and settlement is that of the moisture content. Laboratory consolidation tests are performed under saturated conditions, while such conditions seldom or never exist in the foundation soils. There has been extensive research by D.G. Fredlund on unsaturated soils.

Consultants should not attempt to use the consolidation test results as the basis of the settlement figures presented to their clients. In Terzaghi's words, "The result of soils tests gradually close up the gaps in knowledge and if necessary the designer should modify the design during construction." The "gaps" are the knowledge required as a whole for the design. Indeed, there is a very wide variation in soil and a vast range of natural field conditions.

5.2.3 Direct Shear Test

The direct shear test (Figure 5.4) is the earliest method for testing soil shearing strength. The Box Shear apparatus consists of a rectangular box with a top that can

FIGURE 5.4 Direct Shear Test.

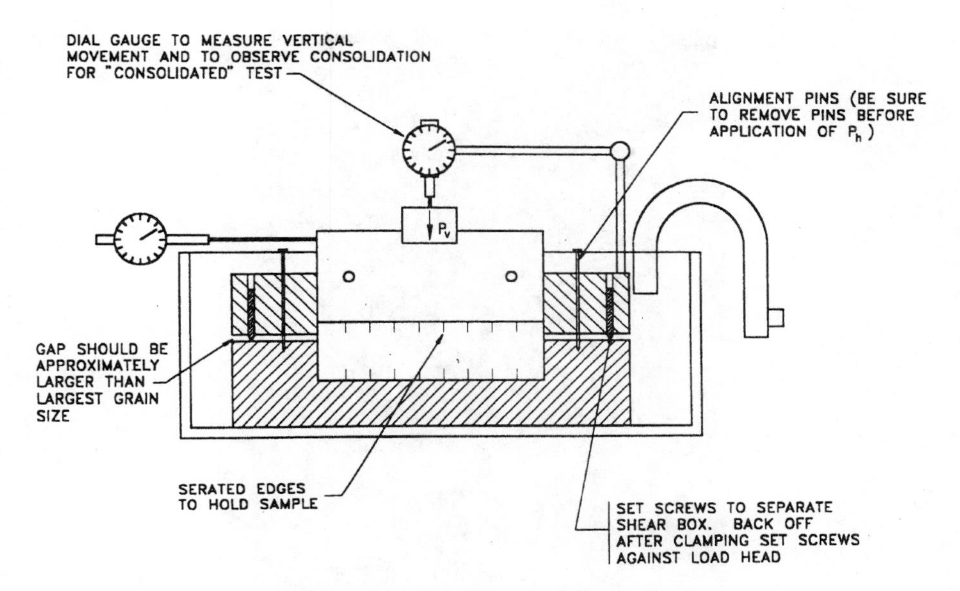

FIGURE 5.5 Direct shear apparatus (after Liu).

slide over the bottom half. Normal load is applied vertically at the top of the box as shown in Figure 5.5.

A shearing force is applied to the top half of the box, shearing the sample along the horizontal surface, and the shear stress that produces the shear failure is recorded. The operation is repeated several times under different normal loads. The resulting values of shearing strength against normal loads are plotted and the angle of internal friction and cohesion value determined.

The commonly used sample size for the direct shear test is 4 in. × 4 in. Such an undisturbed sample can be obtained by the use of a 6-in. Shelby tube. There is an unequal distribution of stresses over the shear surface. The stress is greater at the edges and less at the center. The strength indicated by the test will often be too low. Irrespective of the many shortcomings of the direct shear test, its simplicity led to wide adoption of the test by most consulting engineers' laboratories.

5.2.4 Triaxial Shear Test

The most reliable shear test is the triaxial direct stress test (Figure 5.6). A cylindrical soil sample with a length of at least twice its diameter is wrapped in a rubber membrane and placed in a triaxial chamber. A specific lateral pressure is applied by means of water within the chamber. A vertical load is then applied at the top of the sample and steadily increased until the sample fails in shear along a diagonal plane. The Mohr circles of failure stresses for a series of such tests using different values of confining pressure are plotted as shown in Figure 5.7.

FIGURE 5.6 Triaxial Shear Test.

The advantages of the triaxial shear test over the direct shear test are as follows:

1. The stress is uniformly distributed on the failure plane
2. Soil is free to fail on the weakest surface
3. Water can be drained from the soil during the test to simulate actual conditions in the field
4. A small-diameter sample can be used and the sample preparation is easy.

Triaxial shear test apparatus is costly. Most consulting engineers cannot afford the up-to-date computerized readout. For most foundation investigations, the use of triaxial shear tests are not justified. The bearing pressure values can be obtained from the interpretation of the results of the unconfined compression test. Only in major projects such as earth dam construction, where the values of angle of internal friction and cohesion are critical, should the triaxial shear test be conducted. Many clients today consider the triaxial shear test the apex of soil investigation, and no soil report is complete without such data. It is best for the newly established consulting firms to spend their money on field equipment rather than on the triaxial shear apparatus.

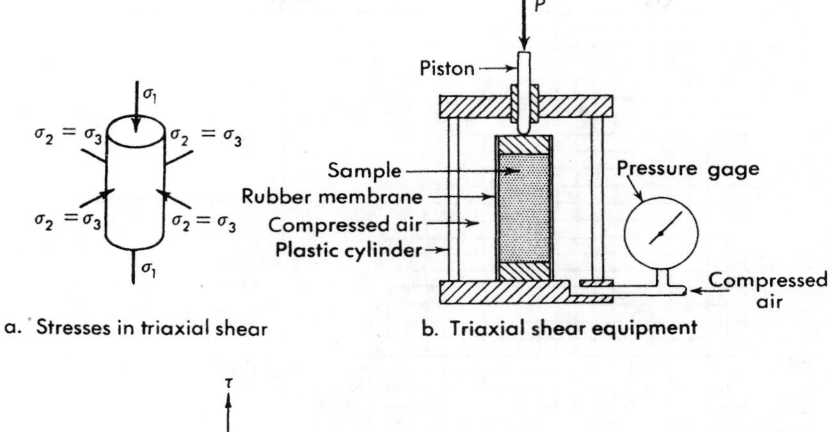

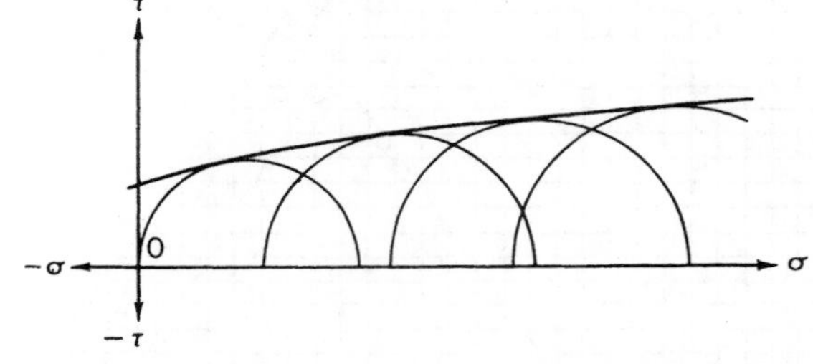

FIGURE 5.7 Triaxial Shear Test (after Sower).

5.2.5 Compaction Test

In 1933, R. R. Proctor showed that the dry density of a soil obtained by a given compactive effort depends on the amount of moisture the soil contains during compaction. For a given soil and a given compactive effort, there is one moisture content called "optimum moisture content" that occurs in a maximum dry density of the soil. Those moisture contents both greater and smaller than the optimum value will result in dry density less than the maximum. A typical compaction curve used by consultants is shown in Figure 5.8.

All geotechnical consultants are familiar with the Proctor Density procedure. The test methods are also listed with ASTM. Essentially, the Standard Proctor Density test consists of compacting the soil into a standard-size mold in three equal layers with a hammer that delivers 25 blows to each layer. The hammer weighs 5.5 lb, with a drop of 12 in. Sixty years after Proctor, his testing procedures are still closely followed, with only minimal refinements. Most laboratories have used mechanical compaction devices to replace hand compaction.

REFERENCES

F.H. Chen, *Foundations on Expansive Soils,* Elsevier, New York, 1988.

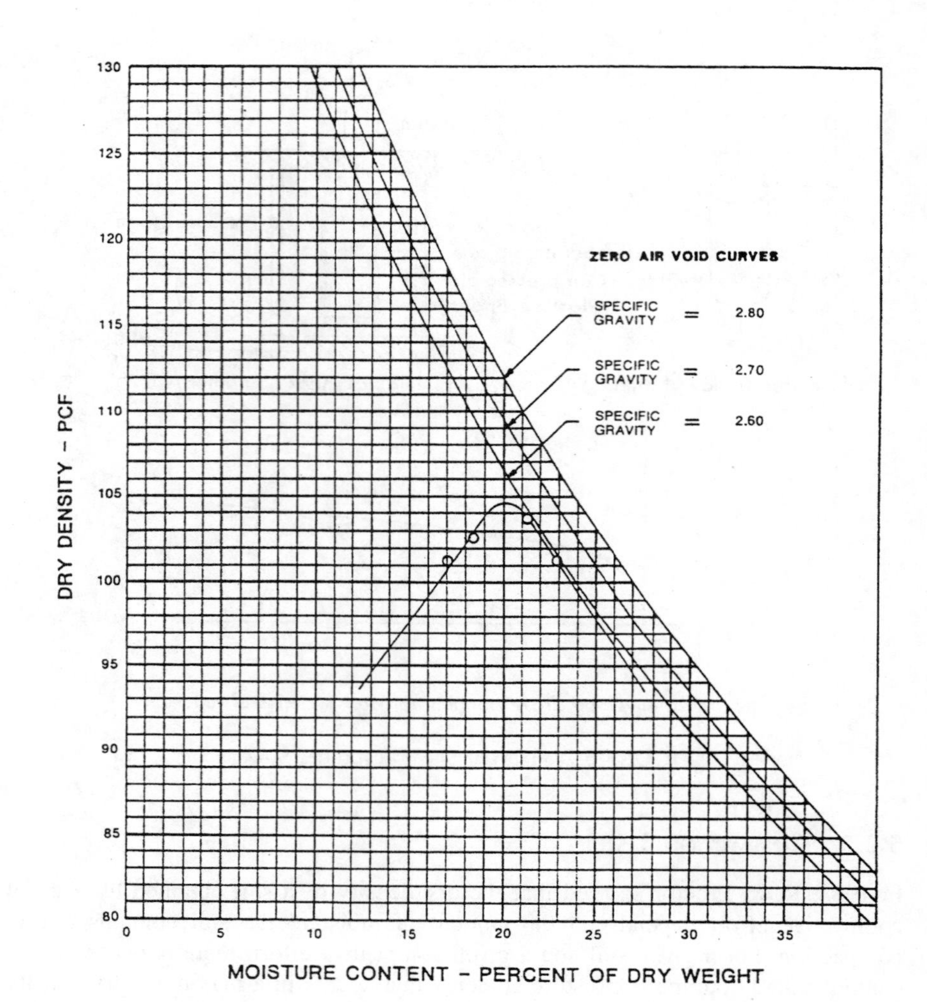

LOCATION : Meadows Boulevard, Station 240+00			MOISTURE-DENSITY RELATIONSHIPS	
HOLE NO. :	DEPTH :	SAMPLE NO. :		
SOIL DESCRIPTION : Sandy Clay			Chen & Associates	
MAX. DRY DENSITY : 104.7PCF		OPT. MOIST. CONTENT : 20.2 %	PROCEDURE : ASTM D698-78 Method A	
LIQUID LIMIT : 39		PLASTICITY INDEX : 19	JOB NO. : 1 405 86	FIG. NO.
GRAVEL : %	SAND : %	SILT AND CLAY (-200) : 73 %	DATE : 6-24-86	3

FIGURE 5.8 Typical Proctor Curve.

B.M. Das, *Principles of Geotechnical Engineering,* PWS Publishing, Boston, 1993.
D.G. Fredlund and H. Rahardjo, *Soil Mechanics for Unsaturated Soils,* John Wiley & Sons, New York, 1993.
C. Liu and J. B. Evett, *Soils and Foundations,* Prentice-Hall, Englewood Cliffs, NJ, 1981.
R.R. Proctor, The Design and Construction of Rolled-Earth Dams, Engineering News Record II, 1993.
G.B. Sowers and G.F. Sowers, *Introductory Soil Mechanics and Foundations,* Collier-Macmillan, London, 1970.
K. Terzaghi, R. Peck, and G. Mesri, *Soil Mechanics in Engineering Practice,* John Wiley-Interscience Publication, John Wiley & Sons, New York, 1996.
R. Whitlow, *Basic Soil Mechanics,* Longman Scientific & Technical, Burnt Mill, Harrow, U.K., 1995.

6 Foundation Design

CONTENTS

More than 70% of projects for average geotechnical consultants are related to building foundations. Their projects range from high-rise buildings to bath houses. In the U.S., most owners will not proceed with construction without a soil test. This is not only for the safety and soundness of the structure but also as a safeguard against future lawsuits. For projects such as subdivision development or industrial parks, the entire area may have to be delineated to comply with the subsoil conditions. For wind resistance structures such as towers and walls, water retaining structures such as ponds and reservoirs, the use of different approaches and various report requirements may be needed.

Consulting engineering is a business, no different from any other highly competitive profession. To maintain the client-consultant relationship, the project must be completed satisfactorily within the designated time and within reasonable cost. By failing to do so, the consultant may not have a second chance. The amount of field investigation must not be excessive. It requires only the necessary number of drill holes and reasonable depth to justify the suggested recommendations. The amount of laboratory testing must be sufficient to justify the design criteria.

When the geotechnical engineers are required to deal with special projects, such as dams, canals, off-shore structures, tunneling, high-rise structures, and the like, there must be sufficient time and budget to allow them to make an in-depth study. Detailed reports will be issued, some of which may even be published at a later date. Such projects do not come often and cannot be depended upon for the establishment of a consultant's office.

It should be borne in mind that technology is not the only means by which to stay in business; competitive pricing certainly cannot be ignored.

6.1 SIGNIFICANCE OF TEST RESULTS

The significance of laboratory testing is to justify the recommendations given by the consultants in their report to the client. Too many soil reports include a large number of test data that have no bearing on the body of the soil report. To a layman, the consultant probably will be praised for taking his project seriously by including so many tests. To an experienced engineer, the report indicates that the writer still needs to understand the significance of laboratory testing.

With the exception of the classification test, the main purposes of the laboratory test are to determine the values of "shear" and "consolidation." Sixty years after Terzaghi, enough papers and books have been written on the two subjects to fill many drawers in the filing cabinet. Still, to the many thousands of geotechnical consultants in the world, the significance of shear tests and consolidation tests is not fully realized.

6.1.1 Unconfined Compression Tests

Unconfined compression tests can be performed with a hand-operated compressor or a more refined speed-controlled mechanical compressor. Both undisturbed and remolded samples can be tested. With the California sampler described in Chapter 2, undisturbed samples extruded from the brass thin-wall liners can be placed directly on the compressor without trimming. The operation is considered the simplest test in the geotechnical laboratory. However, it is amazing to find the wide range of figures derived from the test results. The consultants wonder what they are going to do with all the test results put in front of them by the laboratory technician.

The value of unconfined compression tests is still controversial. Some claim that the use of such tests is limited to special problems and the results are considered to represent index properties rather than engineering properties.

Table 6.1 indicates the relationship between the qualitative terms describing consistency and the quantitative values of unconfined compressive strength. It is obvious such values can only be used as a guide for foundation design.

By accumulating hundreds of unconfined compressive strength data results with their corresponding penetration resistance as shown in Figure 6.1 a meaningful relationship can be found. Figure 6.1 indicates the upper, the lower, and the average values.

In most consultants' offices, there is a great deal of information on the relationships between penetration resistance and unconfined compressive strength value. With such values in a predominately clay soil area, similar curves as shown can be established. By using such a curve, a quantitative unconfined compressive value can be established. Since unconfined compressive strength is really a special case of

TABLE 6.1
Penetration Resistance and Unconfined Compression Strength

Consistency	Field Identification	Unconfined Compressive Strength tons/ft²
Very soft	Easily penetrated several inches by fist	Less than 0.25
Soft	Easily penetrated several inches by thumb	0.25–0.5
Medium	Can be penetrated several inches by thumb with moderate effort	0.5–1.0
Stiff	Readily indented by thumb, but penetrated only with great effort	1.0–2.0
Very stiff	Readily indented by thumbnail	2.0–4.0
Hard	Indented with difficulty by thumbnail	over 4.0

(after Das)

triaxial compression carried out at zero cell pressure, the established value can be used for foundation design. Details are given in Chapter 7.

Without establishing the curve shown in Figure 6.1, the correlation of the unconfined compressive strength value with penetration resistance can be generally expressed as follows:

For average projects, the consistency of clay can be determined by the unconfined compressive strength tests. The reliability of using unconfined compressive strength value for the computation of allowable soil pressure depends on the soil classification. In practice, most clays contain a considerable amount of coarse material with a defined friction angle. Such materials have a higher bearing pressure than plastic clays and at the same time the unconfined compressive strength can be relatively low.

For soft clay, the laboratory's unconfined compressive strength value is not reliable. Its strength depends on the arbitrary standard that the load required to produce the 20% strain is the actual shear failure. The triaxial shear test or the direct shear test should be used for the evaluation of shear strength in soft clay.

For very stiff clays such as weathered or intact claystone shale, the unconfined compressive strength value depends a great deal on the condition of the sample. Drive samples destroy, to a large extent, the structural strength. Generally, only about 30% of its actual strength is shown in the test. Far better samples can be obtained by coring through the shale. However, due to the presence of fissures and slickensides of the sample, the strength value of the shale core sample is probably less than 75% of its actual value.

A rule of thumb is to divide the penetration resistance by two, and the allowable bearing value in kips per square foot is obtained. A detailed discussion is found in Chapter 10, *Pier Foundations*.

6.1.2 Consolidation Tests

The compressibility characteristics of a soil relating to both the amount and rate of settlement are usually determined from the one-dimensional consolidation test or

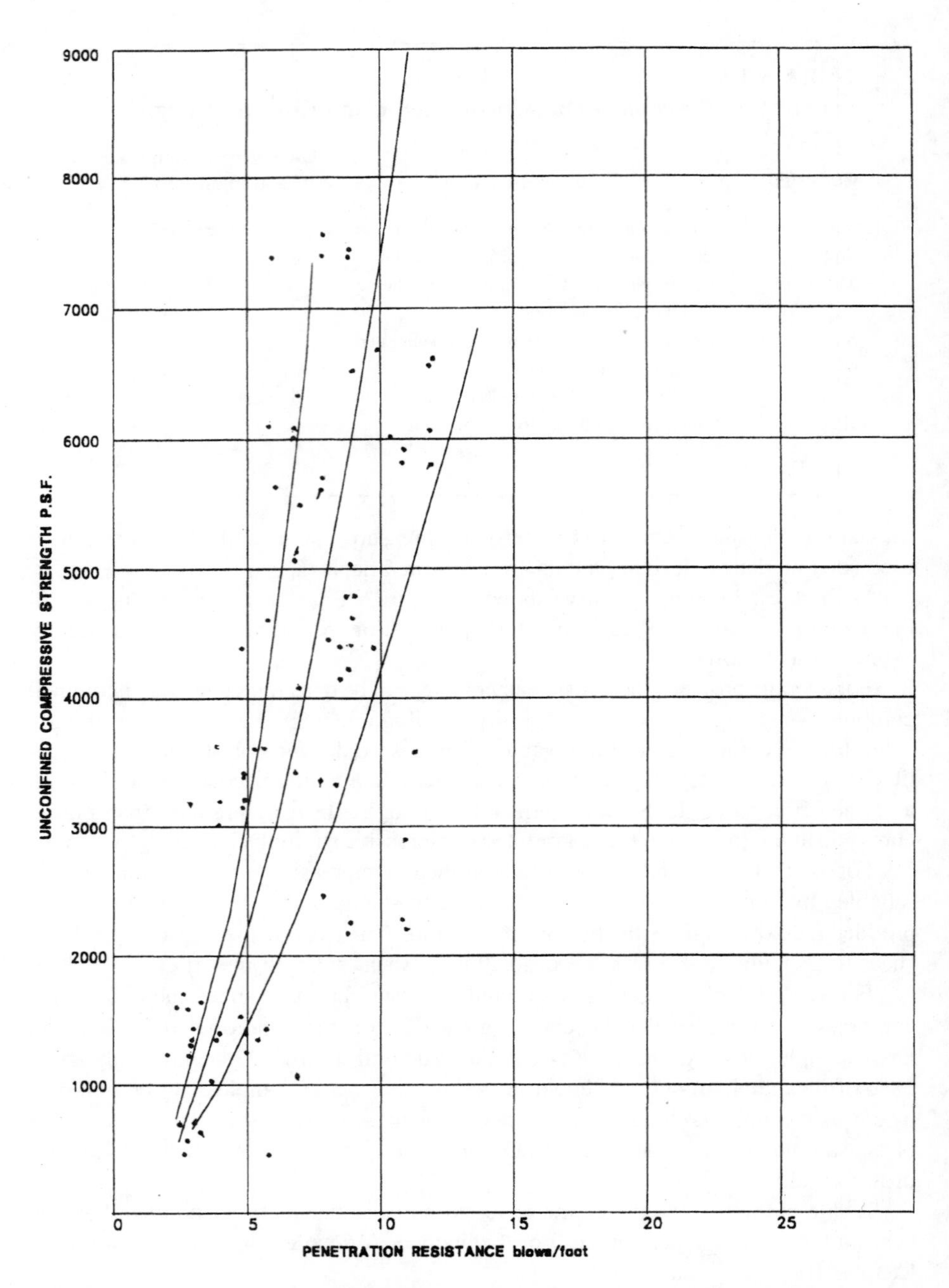

FIGURE 6.1 Approximate relation between penetration resistance and unconfined compressive strength, based on accumulated data in the consulting engineer's office.

TABLE 6.2
Penetration Resistance and Unconfined Compressive Strength of Clay

Consistency	Penetration Resistance Blow/foot	Unconfined Compressive Strength psf
Very soft	0–2	300–1300
Medium	4–8	1300–4000
Stiff	8–15	4000–8000
Very stiff	15–30	8000–15000

(after Peck)

the oedometer test. In a classic theory of consolidation developed by Terzaghi in 1919, a layer of clay was sandwiched between free draining granular soils. Such conditions seldom or never exist in reality. After Terzaghi, the studies of consolidation were extensively reviewed by many leading geotechnical authorities. Many hundreds of papers were published on this subject, some of which address the following topics:

Primary consolidation
Secondary consolidation (creep)
Coefficient of permeability
Initial stress condition
Field consolidation
Over consolidation
Three-dimensional consolidation
Initial stress condition
Rate of consolidation

In practice, consolidation test results can be used as an indicator of the behavior of the structure under load on a short- or long-term basis. Consolidation tests are usually conducted with samples in the in situ moisture content. A typical consolidation curve for medium stiff sandy clay is shown in Figure 6.2. From the curve, the initial concern of a geotechnical engineer is to determine whether the curve indicates nominal consolidation, excessive consolidation, collapse, or expansion. In addition, the following is observed:

1. Under a pressure of 3000 psf, the soil consolidates 7.5%. The uses of percentage of consolidation instead of void-ratio simplify the calculation.
2. The test was performed with saturation of the sample at 1000 psf. Upon saturation under the same pressure, the soil settled 1%. Consequently, it

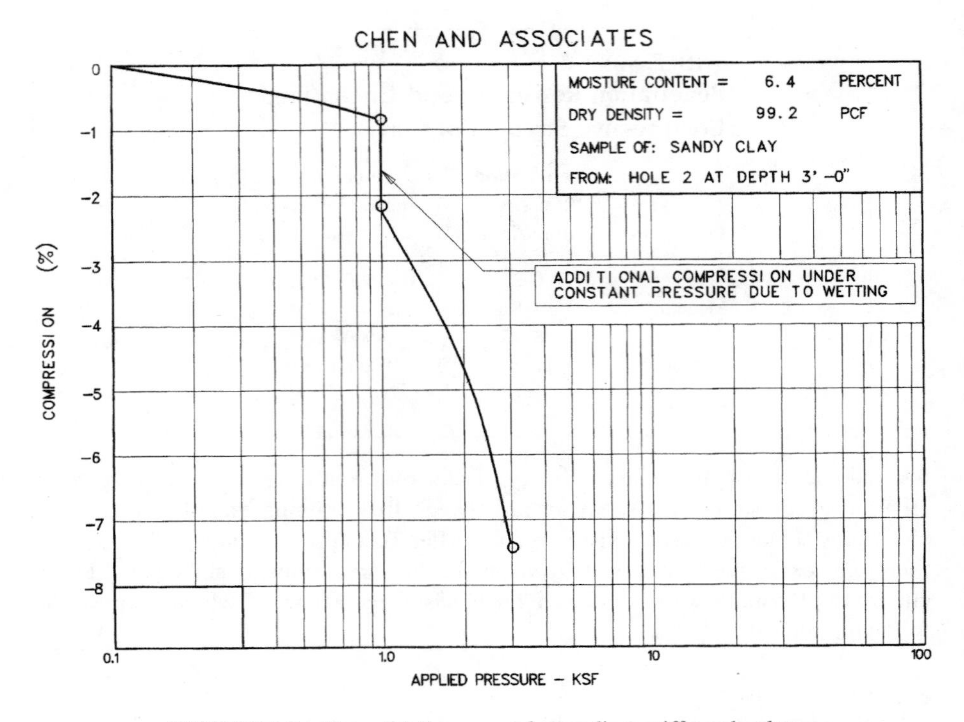

FIGURE 6.2 Consolidation curve for medium stiff sandy clay.

may be estimated that the difference between saturation and in situ condition in consolidation is on the order of 2.5%.

3. A consolidation test on a small sample from 2 to 4 in. diameter cannot reflect the true settlement behavior of the soil under load. The disturbance of the sample during testing contributes to considerably more settlement under laboratory conditions than in actual settlement. This is especially true in the initial phase of the consolidation curve.
4. A conservative method is to assign only 50% of laboratory settlement to actual settlement. In extreme cases, it is believed that only one quarter of the laboratory consolidation actually takes place in the field.

6.2 DESIGN LOAD

To a structural engineer, the load imposed on a structure is obvious. All handbooks clearly define dead load, live load, wind load, and sometimes snow load. The conditions are not always obvious to a geotechnical engineer. Often, at the time the client orders the soil test, the type of structure has not been decided. No information on the design load can be obtained, since the structural engineer has not been selected. Sometimes a client says he is going to build a simple office building, but in fact he is planning a multistory office complex. He thinks that by minimizing the size of his project, his consulting fee will be less.

6.2.1 Dead and Live Load

Dead load refers to the portion of load permanently attached to the structure. Such load essentially is the weight of the structure, including floor finish, walls, ceiling, building frame, and interior finishing. Sometimes, the weights of the footings are not included in the total dead load, on the assumption that the weight of the soil removed during foundation excavation will offset the weight of the footings.

Also included in the dead load calculation is the weight of the earth fill. For instance, the earth cover of a buried tank should be included, although such load can be removed. The weight of water in a water tank sometimes can be considered as dead load, since only in rare cases will the tank be empty.

Earth pressure acting permanently against the portion of structure below the ground surface should be considered as dead load, although some structural engineers consider earth pressure only in their stability analysis.

Dead load evaluation is especially important in the design of piers in an expansive soil area. Dead load is the only element acting on the pier that is not affected by a change of conditions.

Live load includes all loads that are not a permanent part of the structure but are expected to be superimposed on the structure during a part or all of its useful life. For all practical purposes, live load includes human occupancy, furniture, warehouse goods, etc. Also included are snow load, wind load, and seismic load. The building code requires a live load of 40 psf for human occupancy, a snow load of 30 psf to as much as 100 psf, and wind load on the order of one third of snow load. Code requirements vary from city to city.

No rigid definition should be given for dead or live load; the structural engineer should use his discretion in deciding the design criteria. Close communication between the geotechnical engineer and the structural engineer is badly needed. For minor projects, the structural engineer may not even read the soil report. A set of structural drawings is usually absent from the construction office. It is important that the geotechnical engineers should have good knowledge in structural design so that the design criteria given in the report is not confused, making the structural design difficult.

6.2.2 Balanced Design

The characteristics of the soils beneath the footings sometimes determine the design load. To proportion footings on granular soil for equal settlement, the engineer uses the most realistic possible estimate of the maximum live loads rather than arbitrarily inflated ones.

Theoretically, saturated clays will not experience settlement if water is not allowed to escape. A short duration load increment will not be of sufficient magnitude to trigger undue settlement. Because of the slow response of clay to load increment, the settlement should be estimated on the basis of the dead load plus the best possible estimate of the long term average, instead of the maximum live load.

When soil conditions indicate that 1 in. or more of maximum settlement may be expected, the footings for the main structures will be proportioned for uniform dead load pressure. The design of dead load pressure will be determined by using the footing that has the largest ratio of live load to total load. Thus, this particular footing is sized for:

Area = (Total load/allowable soil pressure) and then

Dead load pressure = (dead load/required area).

This dead load pressure is then used to determine the area required for all other footings:

Area = (dead load)/(dead load unit pressure).

The importance of a balanced design for foundations has been overemphasized. For complex structural systems, it requires great effort to balance each footing. The subsoil condition under each footing is more or less different both laterally and in depth. It is only in very unusual cases that one will find the subsoil under each footing identical to that revealed in the drill log. In most cases, it is recommended to decrease the subsoil bearing capacity to cover the effect of unbalanced footing pressure. To a geotechnical consultant who assigns a conservative value to the footings, a balanced design can be a minor consideration.

6.3 SETTLEMENT

To a layman, an architect, an owner, and an attorney, the first concern about a structure is settlement. Would settlement cause damage? Would settlement cause cracking? Would settlement devalue the property? What is the legal responsibility of the geotechnical engineer on settlement? Unfortunately, there is no positive answer. Unlike metals, the amount of elongation or shortening under certain forces can be determined, but the amount of settlement from a structure under load cannot be accurately determined. Highway engineers know how much should be provided in an expansion joint. Concrete engineers know the amount of deflection of the beam. The amount of flow through a conduit can be determined with accuracy. Still, it is a different story when dealing with structures founded on certain soils.

Ever since Terzaghi advanced the "Theory of Consolidation," the theory of settlement has been a favorite subject to geotechnical engineers. More than 20% of all the geotechnical papers published deal with this subject. With today's knowledge of computer science, some claim that they are able to predict settlement to the nearest inch. With the possession of tools such as finite element analysis, plastic flow theory, and others, can the amount of settlement be determined within a reasonable limit?

6.3.1 Permissible Settlement

Structures founded on soils will experience settlement. The magnitude of permissible settlement depends on the type of structure and its function. Uniform settlement seldom presents any serious problems. Foundations founded on granular soils generally complete 75% of their settlement during construction, and consequently, unless accurate elevation readings are taken in the course of construction, such settlement is seldom noticed.

At the same time, drilled pier or pile foundations do experience a considerable amount of settlement in contrast to the belief that foundations founded on bedrock will not settle. The First National Bank building in Denver is founded with piers drilled into the blue shale designed for a maximum soil pressure of 60,000 pounds per square foot with a column load of 3000 kips. Settlement measured shortly after occupation was 0.75 to 1 in.

Structures such as water tanks and silos will settle considerably more than narrow footings, where the depth of pressure bulbs is limited. Settlement of tanks in excess of 6 in. is not uncommon.

Uniform foundation settlement is of little concern to the geotechnical engineer. In fact, many important structures are designed for settlement. At the recently completed Tokyo International Airport, where structural fill extended into the sea, settlement of structures and runway was measured in meters. High-rise buildings in Shanghai founded with raft foundations were designed to accommodate large settlement. Unfortunately, unless very homogenous conditions exist (under uniform wetting and with uniform loading), uniform settlement cannot be expected. In most cases, excessive settlement is incorporated with large differential movement that cannot be tolerated. Structures in Mexico City would not have been the focus of such concern to the public had the settlement been uniform.

6.3.2 Differential Settlement

The amount of differential settlement that can be tolerated by a structure depends on many factors, such as the type of structure, the column spacing, and whether the structure is tied in with existing buildings. Simple span frames can tolerate greater distortion than rigid frames. A fixed-end arch is very sensitive to abutment settlement. Pier or pile settlement in a continuous girder or truss may be critical.

A direct result of differential settlement is cracking. Settlement cracks generally assume a near 45° angle and take place invariably over and below doors and windows. The crack is generally wider at the top than at the bottom. Temperature cracks can sometimes be mistaken for settlement cracks.

Building material has a great deal to do with the extent of cracking. Concrete walls and panels can withstand considerable movement without exhibiting large cracks. Cracks in concrete usually assume a hairline pattern and cannot be detected without close examination. A masonry wall will readily show diagonal cracks and is a good barometer for indication of foundation movement. Cinder block walls cannot tolerate even a very small movement and will crack with temperature changes.

The term "differential settlement" is not clearly defined. The difference of settlement between two adjacent columns is commonly referred to as "differential settlement." Many geotechnical engineers refer to differential settlement as the difference in elevation across the building boundary. For building additions, the difference in elevation between the new addition and the existing building can be totally differential. In describing the amount of differential settlement, the conditions and definitions should be clearly stated.

TABLE 6.3
Commonly Accepted Amount of Settlement

Type of structure	Allowable maximum settlement (in)
Commercial/institutional buildings	1
Industrial buildings	1½
Warehouses	2
Special machinery foundations	As required by manufacturer (often less than 0.002 in.)

(after Teng)

The plaintiff's attorney can present a case to the jury by stating that the disputed project has as much as 6 in. of differential settlement. In fact, the figure refers to the differential from one end of the building to the other, which is within the allowable limit.

6.3.3 Reasonable Settlement

Table 6.3 lists the commonly accepted amount of settlement, as assumed by Teng in 1940.

Generally, it is assumed that differential settlement is about one quarter of the maximum settlement. Architects or owners may have the false impression that a properly designed foundation will not settle. They may be greatly alarmed when they discover that the building moves.

It is the duty of the geotechnical consultant to study the possibility of excessive settlement for the structure of the owner. He or she should communicate with the structural engineer; study the outcome of flooding; investigate the conditions of the neighboring structures; look into the watering practice of the owner; and be aware that the owner may add to the structure in both lateral and vertical directions.

In the report to the owner, all possibilities should be covered. In no case should such sentences as "The amount of settlement will not exceed …" or "The amount of settlement will be on the order of …" be included. Bear in mind that soil mechanics is an "art" and the geotechnical engineer cannot control conditions during or after construction.

A typical lawsuit involved a geotechnical company that estimated the amount of settlement of a structure in its report. When the structure settled more than the estimated amount, it became the focal point of the litigation. Millions of dollars were paid by the consulting firm to settle the case. It is always easy after the fact to point out what should have been done and what should not have been used.

6.4 HEAVE PREDICTION

Expansive soils cover more than one third of the earth's surface. It has only been in the last 50 years that geotechnical engineers have studied its effects on structures.

A great deal of attention has been focused on heave prediction. It is assumed that the amount of heave expected from a structure founded on expansive soils can be predicted in the same pattern as the settlement prediction. It did not take long for the engineers to realize that the problem is much more complex. In the first place, the amount of heave is not merely governed by the loaded depths of the footings. The amount of heave for structures founded on expansive soils depends on many factors that cannot be readily determined. The following basic factors should be considered.

6.4.1 Environmental Change

Climatic conditions involving precipitation, evaporation, and transpiration affect the moisture in the soil. The depth and degree of desiccation affect the amount of swell in a given soil horizon. Considerations should also be given to the drainage problems during the rainy season.

The thickness of the expansive soils stratum extends down to a great depth. The practical thickness is governed by the capability of surface water penetrating into the stratum. It is generally assumed to be a depth of 15 ft, and in some extreme cases, surface water penetrates through the seams and fissures of expansive clays to as much as 30 ft.

6.4.2 Water Table

Soil below the water table should be in a state of complete saturation. Theoretically, no swelling of the clays should take place for the portion of soil below the water table. However, in the case of claystone bedrock, due to the fine-grained structure and the nearly impermeable nature, water is not able to penetrate and saturate the material. Under a fairly stable and unchanging environment, the claystone portion of bedrock below the water table can be considered to be free of volume change. Prediction of total heave can be considered only on the stratum thickness above the water table.

A change of environment can alter the entire picture. Construction operations such as pier drilling can break through the system of interlaced seams and fissures in the claystone structure, thus allowing water to saturate the otherwise dry area, resulting in further swelling of the otherwise stable material.

In considering the thickness of expansive stratum, consideration should be given not only to the thickness above the water table, but also to some depth below the water table. This depth should be at least equal to that of the possible fluctuations of the water table.

6.4.3 Excavation

The predicted amount of heave depends on the initial condition of the soil immediately after construction. If the excavation is allowed to be exposed for a long period of time, desiccation will take place, and upon subsequent wetting, more swelling may take place.

Upon completion of excavation, the stress condition in the soil mass will undergo changes. There will be elastic rebound. Stress releases increase the void-ratio and alter the density. Such physical changes will not take place instantaneously. If construction proceeds without delay, the structural load will compensate for the stress release. Thus, this will not be a significant amount.

6.4.4 Permeability

The permeability of the soil determines the rate of ingress of water into the soil, either by gravitational flow or by diffusion, and these in turn determine the rate of heave. The higher the rate of heave, the more quickly the soil will respond to any changes in the environmental conditions, and thus the effect of any local influence is emphasized. At the same time, the higher the permeability, the greater the depth to which any localized moisture will penetrate, thus engendering greater movement and greater differential movement. Therefore, the higher the permeability, the greater the probability of differential movement.

6.4.5 Extraneous Influence

The above-mentioned basic factors, although difficult to predict, can be evaluated theoretically. At the same time, extraneous influences are totally unpredictable. The supply of additional moisture will accelerate heave, for instance, if there is an interruption of the subdrain system to allow the sudden rise of a perched water table. The development of the area, especially residential construction, can contribute to a drastic rise of the perched water table.

Various methods have been proposed to predict the amount of total heave under a given structural load. These include the double oedometer method, the Department of Navy method, the South Africa method, and the Del Fredlund method. Recently, with the advance of suction study, Johnson and Snethen claimed that the suction method is simple, economical, expedient, and capable of simulating field conditions.

Some fundamental differences between the behavior of settling and heaving soil are as follows:

1. Settlement of clay under load can take place without the aid of wetting, while expansion of clay cannot be realized without moisture increase.
2. The total amount of heave depends on the environmental conditions, such as the extent of wetting, the duration of wetting, and the pattern of moisture migration. Such variables cannot be ascertained, and consequently, any total heave prediction can only be speculation.
3. Differential settlement is usually described as a percentage of the ultimate settlement. In the case of swelling soils, one corner of the structure may be subject to maximum heave due to excessive wetting, while another corner may have no movement. No correlation between differential and total heave can be established.

6.5 BUILDING ADDITIONS

Take great care when designing a new addition adjacent to or abutting an existing building. This is especially important when the existing structure is owned by another person. The new footings can exert an additional load on the existing footings and cause settlement and cracking. Whenever possible, it is wise to consult with the original engineer or the owner and study the initial design. If common walls are used, eccentric loading will be expected. When the new and the old structures are not on the same level, the lateral load from the existing structure should be considered. The bearing capacity as calculated for isolated footings should be drastically reduced.

Similar precautions should be taken even when the new construction is isolated from the existing structure. The owner of the neighboring structure can claim that the weight of the new construction has caused the settlement of the neighboring structure. It is therefore important to have a conference with the neighboring building owners before starting the excavation. A prudent engineer takes pictures of the neighboring structure to avoid possible future litigation. Documented photographs can prove that the distress or cracking of the neighboring building existed before the new construction.

Another important consideration in the design of footings is the property line. The building owner wants to make use of every foot of his property. Without the knowledge of the adjacent property owner, the footing construction may extend beyond the property line. The error may not be detected until years later when the excavation of the neighboring property is started. The court can order the demolition of the building or order the payment of a substantial compensation.

It is very rare for a geotechnical consultant to be sued for overdesign, but neglecting to pay attention to the site condition can haunt the engineer. Details such as neighboring structures, property lines, drainage patterns, slope stability, or the rise of water table may be more important than the accuracy of the bearing capacity numbers.

REFERENCES

F.H. Chen, *Foundations on Expansive Soils,* Elsevier Science, New York, 1988.

B.M. Das, *Principles of Geotechnical Engineering,* PWS Publishing, Boston, 1994.

P. Rainger, *Movement Control in Fabric of Buildings,* Batsford Academic and Educational, London, 1983.

D. R. Sneathen and L. D. Johnsion, Evaluation of Soil Suction from Filter Paper, U.S. Army Engineers, Waterway Experimental Station, Vicksburg, Mississippi, 1980.

W.C. Teng, *Foundation Design,* Prentice-Hall, Englewood Cliffs, NJ, 1962.

K. Terzaghi, R. Peck, and G. Mesri, *Soil Mechanics in Engineering Practice,* John Wiley-Interscience Publication, John Wiley & Sons, New York, 1996.

U.S. Department of the Interior, Bureau of Reclamation, *Soil Manual,* Washington, D.C., 1970.

R. Weingardt, All Building Moves — Design for it, *Consulting Engineers*, New York, 1984.

7 Footings on Clay

CONTENTS

The design of footings on clay has been the concern of engineers since the beginning of soil engineering. The classical theory of ultimate bearing capacity developed by Terzaghi more than 60 years ago is still the basic theory used by engineers. In referring to footings on clay, the correct description should be footings on fine-grained soils. These include lean clay, fat clay, and plastic silt; the analysis can sometimes be extended to clayey sands (SC) and sandy silt (ML). The basic requirements of designing footings on clay are that the design should be safe against shear failure and the amount of settlement should be tolerable. The shear consideration is theoretically important; it seldom takes place in actual construction. When such failure does occur, it receives attention from the public. The silo tilting in Canada certainly is a good example.

Consultants are generally conservative and the cost of a slightly bigger footing seldom affects the total construction cost. As discussed in the previous chapter, what constitutes a "tolerable settlement" is hard to define. Judgment and experience of the consultant are probably more important than figures and equations.

7.1 ALLOWABLE BEARING CAPACITY

The ultimate bearing capacity is defined as the intensity of bearing pressure at which the supporting ground is expected to fail in shear. The allowable bearing capacity is defined as the bearing pressure that causes either drained or undrained settlement or creep equal to a specified tolerable design limit. In plain consulting engineer's language, allowable bearing capacity refers to the ability of a soil to support or to hold up a foundation and structure.

In 1942, Terzaghi expressed the ultimate bearing capacity of footing on clay with the following general equation:

$$q_{ult} = cN_c + \gamma\, DN_q + 0.5\, \gamma\, BN_\gamma$$

where q_{ult} = ultimate bearing capacity, psf
γ = unit weight of soil, pcf
c = cohesion, psf
D = depth of foundation below ground, ft.
B = width of footing, ft.
N_c, N_q, N_γ = bearing capacity factors.

The bearing capacity factors are shown in Figure 8.2. The third term of the equation refers to the friction of the soil. For clay, where $\phi = 0$, the term is eliminated. The second term of the equation is referred to as the depth factor. It depends on the construction requirement. In probably 90% of the cases, footings are placed at a shallow depth. Therefore, for footings on clay, the net-bearing capacity can generally be defined as the pressure that can be supported at the base of the footings in excess of that at the same level due to the surrounding surcharge.

$$q_d = cN_c$$

where q_d is the net ultimate bearing capacity. Prandtl determined the value of N_c, for a long continuous footing on the surface of the clay deposit where the friction angle is assumed to be zero, as 5.14. A great deal of research has been conducted in recent years on the bearing capacity factors. The ratio between footing width and footing depth appears to be an important controlling factor.

In general geotechnical practice for low rise structures, the footing width is on the order of 24 to 30 in. For frost protection, the building code generally specifies a 30-in. soil cover. Consequently, the D/B ratio is generally less than one, and the N_c value should be on the order of 5.5 to 6.5, as shown in Figure 7.1.

Using a factor of safety of three, the allowable soil bearing pressure q_a for footings on clay would be

$$q_a = \frac{c\, N_c}{3}$$

For $\phi = 0$ or very small, the unconfined compressive strength is twice the cohesion value of clay. Thus,

$$a_a = \frac{N_c\, q_u}{6}$$

$$= q_u$$

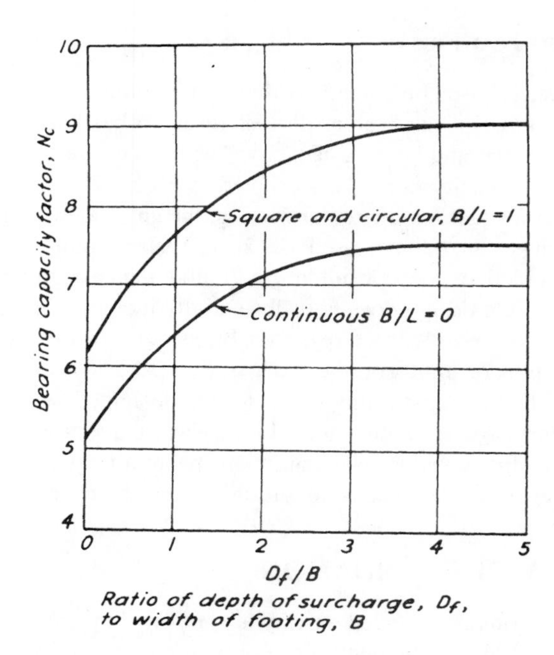

FIGURE 7.1 Bearing capacity factors for foundation on clay (after Skempton).

where q_u is the unconfined compressive strength. For most structures the consultants are dealing with, it will be sufficient to assume that the allowable soil-bearing pressure for footing on clay is equal to the unconfined compressive strength. In using the unconfined compressive strength values for footing designs, the following should be considered:

Average value — It is a mistake to determine the value by averaging all the data obtained from the laboratory. Experience should guide the consultant in selecting the most reliable and applicable ones.

Water table — The vicinity of the water table or the likelihood of the development of a perched water condition should be of prime importance in selecting the design value. Most foundation failures take place, not due to underdesign, but due to the failure to recognize the possibility of the saturation of the footing soils.

Drainage — It is common practice to provide drains along the footings with the intention of keeping the foundation dry. Such drains may not have an adequate outlet, or sometimes the outlet has been blocked. As a result, the soils beneath the footing can be completely saturated for years without detection.

Soft layer — The presence of a soft layer sandwiched between relatively firm clays should not be ignored. During exploratory drilling, such a layer can be overlooked by the field engineer. If such condition is suspected, the bearing capacity should be reduced.

7.1.1 Shape of Footings

The above analysis is based on Terzaghi's theory of continuous footings, a condition that rarely exists in practice. A great deal of research has been conducted on the effects of footing shape and bearing capacity. The ratio between breadth and length affects the bearing capacity factor N_c as shown in Figure 7.1.

In general, for a square or a circular footing, the calculated bearing capacity for continuous footings can be increased by 20%, that is, multiplying by a factor (1+0.2 B/L). In practice, the consultants will find that assigning a conservative bearing capacity to the design does not substantially increase the construction cost. For small or medium-sized structures, it is often not worthwhile to argue about bearing capacity plus or minus on the order of 500 psf

Bear in mind that the controlling factor for the design of footings on clay is the unconfined compressive strength value. In case of a questionable site, the field engineer should be instructed to take continuous penetration tests and samplings, so that any soft layer or any erroneous condition will not be overlooked.

7.2 STABILITY OF FOUNDATION

The stability of a structure founded on clay is controlled by the safety against shear failure and with tolerable settlement. Since only in rare cases does foundation shear failure take place, the design criteria is generally governed by settlement considerations. To estimate the amount of settlement, it is necessary to study the loaded depth of the footings and the consolidation characteristics of the clay.

7.2.1 Loaded Depth

The classical pressure bulb theory based on Boussinesq's equation can be used. The shapes of the pressure bulb for continuous, circular, and square footings are shown in Figure 7.2.

Examination of the stress distribution within the pressure bulb indicates the following:

TABLE 7.1
Stress Distribution Within the Pressure Bulb

Depth Below Footing Width B	Percentage of Uniform Pressure for Square Footing	Percentage of Uniform Pressure for Continuous Footing
0.5 B	70%	80%
1.0 B	35%	55%
1.5 B	18%	40%
2.0 B	12%	28%

The most commonly used pressure bulb is the one for 0.2 q since in practical cases any stress less than 0.2 q is often of little consequence. Therefore, for all

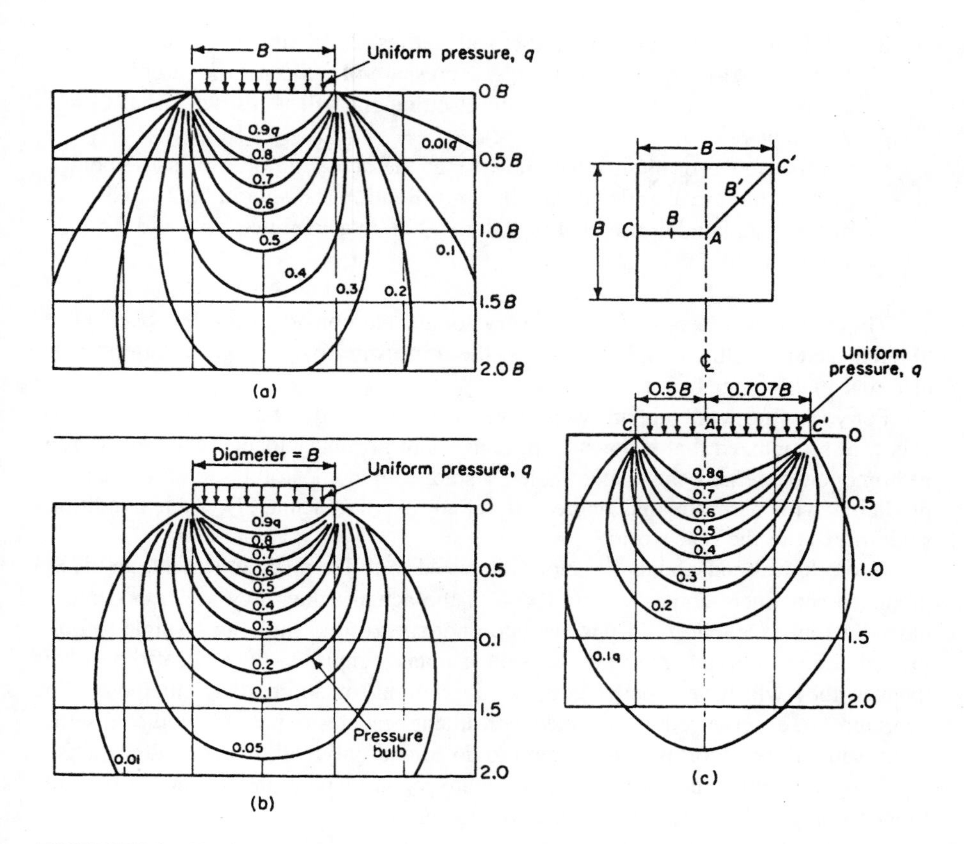

FIGURE 7.2 Vertical stresses under footings: (a) under a continuous footing; (b) under a circular footing; (c) under a square footing.

practical purposes, the pressure bulb for a square footing can be considered as 1.5 B wide and 1.5 B deep, B being the width of the footing.

7.2.2 Consolidation Characteristics

Typical consolidation characteristics of clay are given in Chapter 6 under *Consolidation Test.* Referring again to the consolidation test result as indicated in Figure 6.2, the amount of settlement can be estimated as follows:

1. For a footing width of 30 in., the depth of the pressure bulb according to the theoretical approach is 2.5 times the footing width. Since the effective pressure is only about 80% of the actual pressure, and the effective depth of the pressure bulb is less than the theoretical amount, it is assumed that actual effective depth is only on the order of 1.5 times the footing width.
2. Based on the above assumption, the amount of settlement for a 30-in.-wide footing under a pressure of 3000 pounds per square foot in a saturated condition is (30)(1.5)(7.5%) = 3.4 in.

3. With the in situ condition, the soil settles 2.5% under a pressure of 1000 psf. It is estimated that under a pressure of 3000 psf the sample will settle only 7.5 to 2.5%. The footing settlement will be (30)(1.5)(5.0%) = 2.3 inches.
4. On the above basis, it is estimated that the actual amount of settlement of the structure as reflected by the consolidation test should be 25 to 50% of the calculated figure, that is, 0.8 to 1.7 in. in a saturated state and 0.6 to 1.2 in. in the in situ state.

The above estimate is of course very rough. No consideration has been given to such factors as the sample thickness, the uniformity of the soil, duration of the test, and many other factors.

For years, the academicians were interested in the study of settlement prediction. It is well recognized that if the subsoil consists of normally loaded clay, the subsoil is homogeneous, and the water table is stable, then the total settlement can be predicted with a reasonable degree of reliability. Unfortunately, such conditions seldom exist in the real world.

Geotechnical consultants are more interested in differential settlement, and if the predicted settlement comes within 100% of the actual value, they are considered to have done an excellent job. Consultants do not spend time studying a single sample; instead, they would rather perform tests on as many samples as they can afford. In this manner, they will have a better grasp of the amount of differential settlement to be expected. An experienced geotechnical consultant hesitates to put any predicted settlement value in the report unless required to do so and only with many qualifications.

For geotechnical consultants dealing with recommendations for most structures founded on clay, the following steps are suggested:

1. Assign soil bearing pressure based on penetration resistance and unconfined compressive strength tests for the ultimate value. Select the logical values instead of using the maximum or the minimum values.
2. Check the amount of maximum settlement by consolidation test.
3. Review the assigned value by checking with existing data.

7.3 FOOTINGS ON SOFT OR EXPANSIVE CLAYS

This chapter deals essentially with shallow foundations founded on clay. The structures most geotechnical consultants encounter are small- or medium-sized buildings such as schools, medium-height apartments, warehouses, etc., where elaborate studies are not required or cannot be afforded. Oddly, these are projects that give the consultants the most problems. Lawsuits generated by these owners can often ruin one's business.

At the same time, where sufficient funding is reserved for detailed study, larger projects are highly competitive and seldom acquired. Interestingly, most of the hundreds of papers published in technical journals discuss problems seldom encountered. Soil engineering deeply involved with geology, hydrology, or structures will not be included in this book.

7.3.1 Raft Foundation

A raft foundation is a combined footing that covers the entire area beneath a structure and supports all the walls and columns. A raft foundation is used when the allowable soil pressure is so small that the use of an individual footing will not be economical. A typical example of such a case is the San Francisco area, where the bay mud is soft and the firm bearing stratum deep.

Since the area occupied by the raft is limited by the area occupied by the building, it is difficult to change the soil pressure by adjusting the size of the raft. The design of a raft foundation should be a joint effort between the structural engineer and the geotechnical engineer. Since the loaded depth of a raft does not control settlement, the depth at which the raft is located is sometimes made so great that the weight of the structure is compensated for the weight of the excavated soil.

If very soft clay is encountered and it is necessary to place the footings on such clay, careful analysis of the shear strength of the clay is necessary. The use of a vane shear test correctly interpreted presents the most reliable results. The triaxial shear test is time-consuming and its results depend a great deal on the selected procedure. An experienced operator is necessary to render accurate results. The direct shear test is simple, requiring less operation skill. Unit cohesion obtained from the direct shear test is sometimes more reliable than the unconfined compression test.

7.3.2 Footings on Expansive Soils

The design of footings on expansive soils did not receive attention until recent years. This is probably because much of the expansive soil is located in arid, underdeveloped areas.

Contrary to settlement, expansive soils heave upon wetting. The design criteria for footings on expansive clay is not focused on the allowable bearing pressure but on the swelling pressure. The swelling pressure of expansive soils can exceed 15 tons per square foot. For footing design, the following basic factors should enter into consideration:

1. Sufficient dead load pressure should be exerted on the footings to balance the swelling pressure.
2. The structure should be rigid enough so that differential heaving can be tolerated.
3. The swelling potential of the foundation soils can be eliminated or reduced.

7.3.3 Continuous Footings

Instead of using wide footings to distribute the foundation load, footings on expansive clays should be as narrow as possible. The use of such construction should be limited to clays with a swelling potential of less than 1% and a swelling pressure of less than 3000 pounds per square foot. The limiting footing width is the width of the foundation wall. Continuous footings are widely used in China, Israel, Africa, and other parts of the world where the subsoil consists of illite instead of montmorillonite.

7.3.4 Pad Foundation

The pad foundation system consists essentially of a series of individual footing pads placed on the upper soils and spanned by grade beams. The system allows the concentration of the dead load. Thus, the swelling pressure can be balanced. The use of a pad foundation system can be advantageous where the bedrock or bearing stratum is deep and cannot be reached economically with a deep foundation system.

It is theoretically possible to exert any desirable dead load pressure on the soil to prevent swelling. Actually, the capacity of the pad is limited by the allowable bearing capacity of the upper soils. If the pads are placed on stiff swelling clays, the maximum bearing capacity of the pad is limited by the unconfined compressive strength of the clay.

If $q_u = 5000$ psf, the practical dead load pressure that can be applied to the pad is about 3000 psf (assuming the ratio of dead and live loads to be about one to three). With this limitation, the individual pad foundation system can only be used in those areas where the soils possess a medium degree of expansion with a volume change on the order of 1 to 5% and a swelling pressure in the range of 3000 to 5000 psf.

7.3.5 Mat Foundation

Mat foundation is actually a type of raft foundation. Instead of distributing the structural load, it distributes the swelling pressure. The mat should be designed to receive both the positive and the negative moments. Positive moment includes those induced by both the dead and the live load pressures exerted on the mat. Negative moment consists mainly of that pressure caused by the swelling of the under-mat soils. There would be tilting of the mat, but the performance of the building would not be structurally affected. The limitations of such a system are:

1. The system thus far is limited to moderately swelling soils.
2. The configuration of the structure must be relatively simple.
3. The load exerted on the foundation must be light.
4. Single-level construction is required. It would be difficult to apply such construction to buildings with basements.

Mat foundation systems have been widely used in southern Texas, where moderate swelling soils are encountered. The design of a mat foundation should be in the hands of both structural and geotechnical engineers.

REFERENCES

F.H. Chen, *Foundations on Expansive Soils,* Elsevier Science, New York, 1988.

R. Peck, W. Hanson, and T.H. Thornburn, *Foundation Engineering,* John Wiley & Sons, 1953.

A. W. Skempton, The Bearing Capacity of Clays, Proc, British Bldg. Research Congress, 1, 1951.

K. Terzaghi and R. Peck, *Soil Mechanics in Engineering Practice,* John Wiley & Sons, 1945.

K. Terzaghi, R. Peck, and G. Mesri, *Soil Mechanics in Engineering Practice,* John Wiley-Interscience Publication, John Wiley & Sons, 1996.

8 Footings on Sand

CONTENTS

The principle of the design of footings on sands is essentially the same as the design of footing on clays. In soil mechanics, the definition of sand refers to cohesionless soils with little or no fines. This includes gravely sands, silty sands, clean sands, fairly clean sands, and gravel.

Engineers as well as the public generally have the conception that sandy soils are good bearing soils and will not pose much of a foundation problem. In fact, in Riyadh, Saudi Arabia, during the oil boom period, most structures were erected on sandy soils without the benefit of soil investigations.

In fact, distress experienced on structures founded on sand is not uncommon, especially for subsoils containing large amounts of cobbles. Excessive settlement or sometimes even shear failure can take place when there is a sudden change of the water table elevation.

8.1 ALLOWABLE BEARING CAPACITY

The criteria for designing a safe foundation on sand are the same as those for footings on clay. That is, the possibility of footings breaking in the ground generally refers to "shear failure" or "punching shear," and settlement produced by the load should be within a tolerable limit.

For structures founded on sandy soils, the settlement can take place almost immediately. The settlement criteria generally determine the allowable bearing capacity. Still, the possibility of shear failure cannot be ignored. Geotechnical engineers often found that punching shear took place at narrow footings and was ignored in the design.

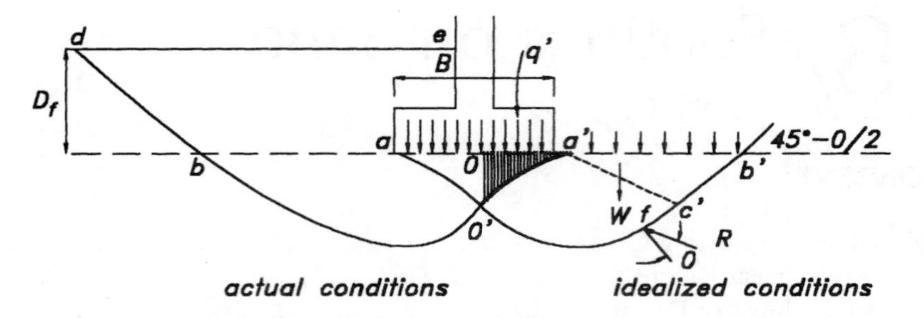

FIGURE 8.1 Cross-section through long footing on sand (left side, after Peck).

Sometimes it is difficult to determine whether the structure failed due to excessive settlement or due to shear. For a consulting engineer dealing with medium or loose sand, an ample factor of safety should be used. The design criteria should depend on the results obtained from in situ testing rather than from theoretical analysis.

8.1.1 Shear Failure

The concept of shear failure of footing on sands (Figure 8.1) was established first by Prandtl and later extended by Terzaghi, Meyernof, Buisman, Casqot, De Beer, and many others.

The general approaches of all studies are similar. They usually follow the basic assumption that the soil is homogenous, from the surface to a depth that is at least twice the width of the footings.

As explained by Peck, the wedge a'o'd cannot penetrate the soil because of the roughness of the base. It moves down as a unit. As it moves, it displaces the adjacent material. Consequently, the sand is subjected to severe shearing distortion and slides outward and upward along the boundary's o'bd. The movement is resisted by the shearing strength of the sand along o'bd and the weight of the sand in the sliding masses.

The mechanics involving the ultimate bearing capacity under such a condition is very complex. It involves the passive pressure exerted by the adjacent soils, further complicated by the drained and undrained conditions. The result in which the consultants are interested is that the ultimate bearing capacity may be expressed as

$$qd' - \frac{1}{2} B\gamma \ N\gamma + \gamma \ D_f \ N_q$$

and the net ultimate bearing capacity as

$$qd = qd' - \gamma \ D_f$$

$$= \frac{1}{2} B_\gamma \ N_\gamma + \gamma \ D_f \left(N_q - 1\right).$$

where $N_\gamma N_q$ are bearing capacity factors, their values can be evaluated by Figure 8.2.

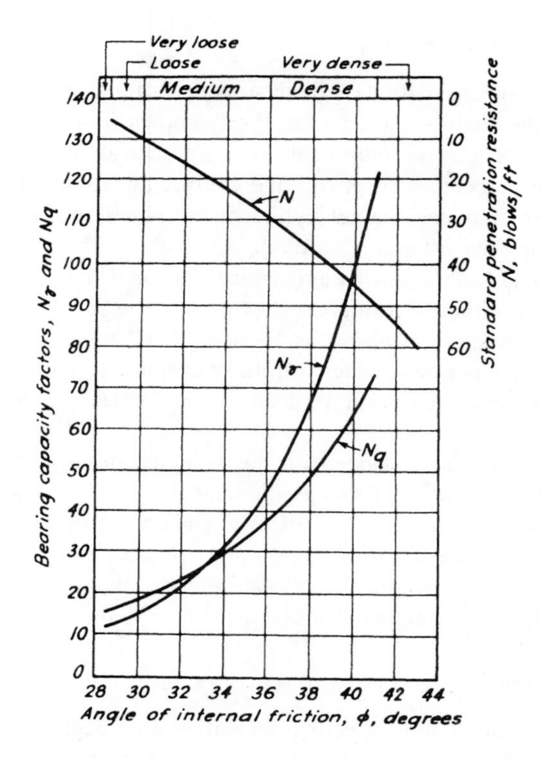

FIGURE 8.2 Relation between bearing capacity factors and angle of internal friction and penetration resistance (after Peck).

8.1.2 Relative Density

The relative density of sand is defined by the equation:

$$Dr = \frac{(e_o - e)}{(e_o - e_{\min})}$$

in which e_o = void ratio of sand in its loosest state
$e_{\min}$ = void ratio of sand in its densest state, which can be obtained in the laboratory
e = void ratio of sand in the field

Relative density can be determined when the maximum, the minimum, and the actual field density of the sand are known. The more uniform the sand (SP), the nearer its e_o and e_{min} will approach the values of equal spheres. For well-graded sands (SW), both e_o and e_{min} values are small and as a result a higher relative density value is expected.

8.1.3 Penetration Resistance

The accuracy in the determination of the in situ angle of internal friction for granular soils depends on the quality of the undisturbed samples. The procedures involve an experienced operator, costly equipment, and time-consuming activities. For soil containing a large percentage of gravel and cobble, laboratory testing depends on the use of a large-diameter triaxial cylinder. For small projects, such elaborate sampling and testing are often not justified.

The simplest and least expensive procedure is to correlate internal friction value with standard penetration test results, as shown in Figure 8.2. The standard penetration test result can often be deceiving, as discussed in Chapter 3. Data obtained from the penetration resistance test, when carefully employed, still presents the most convenient and inexpensive method for determining the bearing capacity of granular soils.

Geotechnical consultants depend on the field penetration resistance value to assign the bearing capacity of footings on sand. Figure 8.3 is based on depth of surcharge D = 3 and a factor of safety of two. The choice of the factor of safety is discussed under a separate heading.

For most projects, the width of the footing is less than 5 ft. The beginnings of the curves actually control most construction.

8.1.4 Gradation

The gradation of granular soils directly affects their relative density, and hence their bearing capacity. When gravely soils are encountered, their bearing value as well as their amount of settlement can be different from sands with the same angle of internal friction. Both the maximum and the minimum densities increase with the higher percentage of gravel, up to probably as much as 60%. The presence of cobbles also greatly affects the bearing capacity. The larger the size of the aggregate, the higher the maximum and the minimum densities.

If the soil contains more than 50% of gravel with a maximum size exceeding 2 in., the values of the bearing capacity assigned for sands as indicated by the previous curves are usually very conservative.

Gradation analysis should be performed for every granular soil. Care should be taken that the sample represents the average site subsoil. With the information furnished from the gradation analysis, it is possible to have a better correlation between the penetration resistance and the bearing capacity.

8.1.5 Meyerhof's Analysis

Bearing capacity analysis made by G.G. Meyerhof assumed that the shear zone extends above the foundation level. Consequently, he assigned a much higher bearing value than that of Terzaghi. Meyerhof estimated the relationship between density, penetration resistance, and angle of internal friction of cohesionless soil as in Table 8.1.

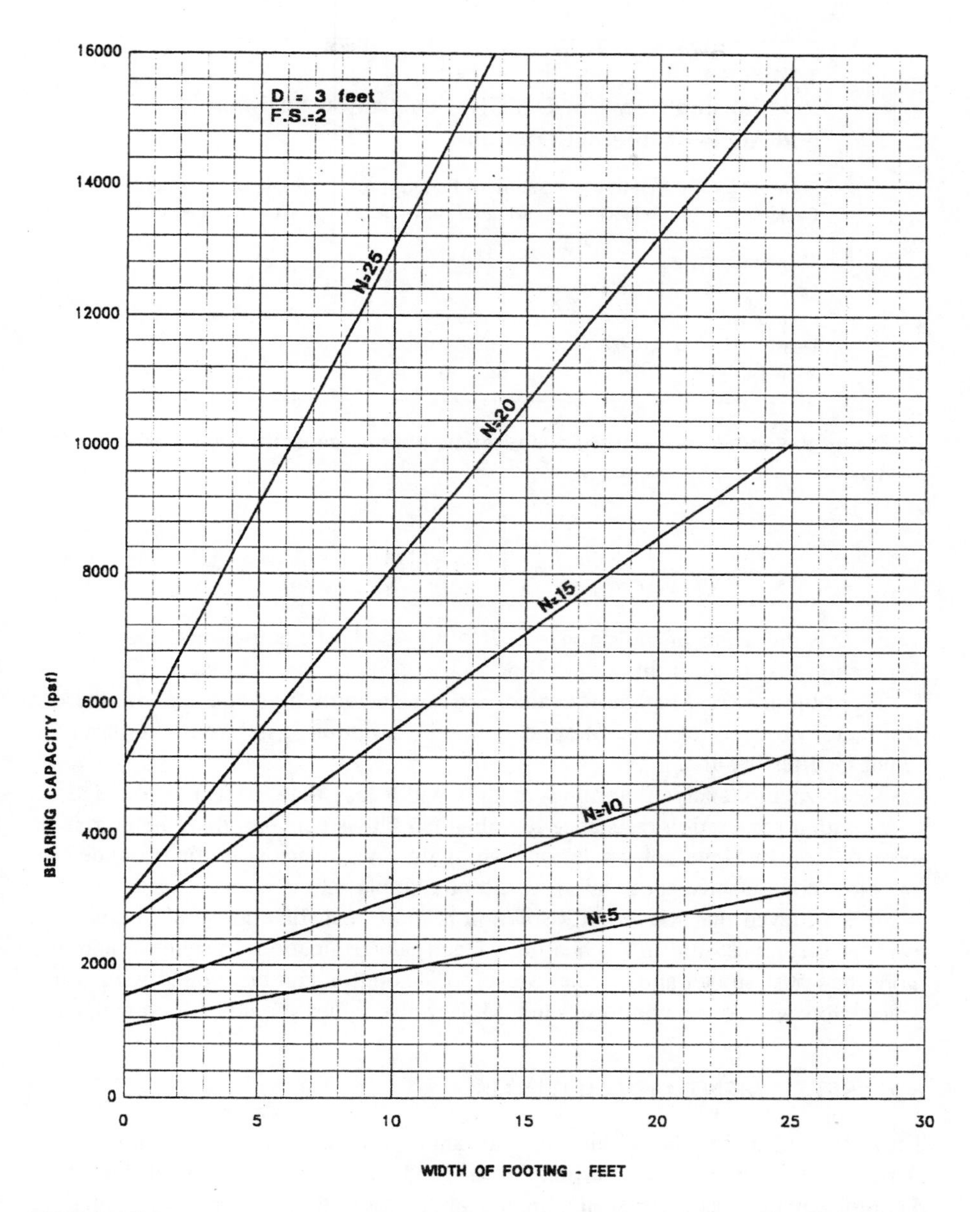

FIGURE 8.3 Width of footing versus bearing capacity for various penetration resistance value.

TABLE 8.1
Relationship Between Density, Penetration Resistance, and Angle of Internal Friction

State of Packing	Relative Density	Standard Penetration Resistance (blow/foot)	Angle of Internal Friction (degree)
Very loose	0.2	4	30
Loose	0.2–0.4	4–10	30–35
Compact	0.4–0.6	10–30	35–40
Dense	0.6–0.8	30–50	40–45
Very dense	0.8	50	45

Using the above relationship, the ultimate bearing capacity can be expressed as follows:

$$q_d = NB\left[1+\frac{D}{(B)}\right]200$$

where D is the depth of the surcharge, B is the width of the footings, and N is the penetration resistance in blows per foot.

From Figure 8.4, Meyerhof's values are considerably higher, even with the use of a factor of safety of three. It appears that his solution is nearer to the observed actual load test results.

In Terzaghi's analysis, footings with a width less than 10 ft and founded on sands with a penetration resistance less than ten blows per foot, there is the risk of shear failure. In Meyerhof's solution, even with a very narrow footing founded on low blow count sands, shear failure will not take place.

It is believed that Meyerhof's solution is more realistic and should be used at least for the upper limit in the bearing capacity determination. In sands containing more than 50% gravel and cobble, Meyerhof's solution can be applied with confidence. However, for fine uniform sand, Meyerhof's values should be used with care.

8.2 SETTLEMENT OF FOOTINGS

The stress and strain relationships of sand cannot be approximated by a straight line. Hence, the term modulus of elasticity of the sand mass cannot be applied. An elastic theorem cannot be used for estimating the amount of settlement for footings on sand under a static load. The factors affecting the settlement of footings on sand are as follows:

1. **Relative Density** — The rigidity of a sand mass increases sharply with the increase of its relative density
2. The shape and size of the sand grain

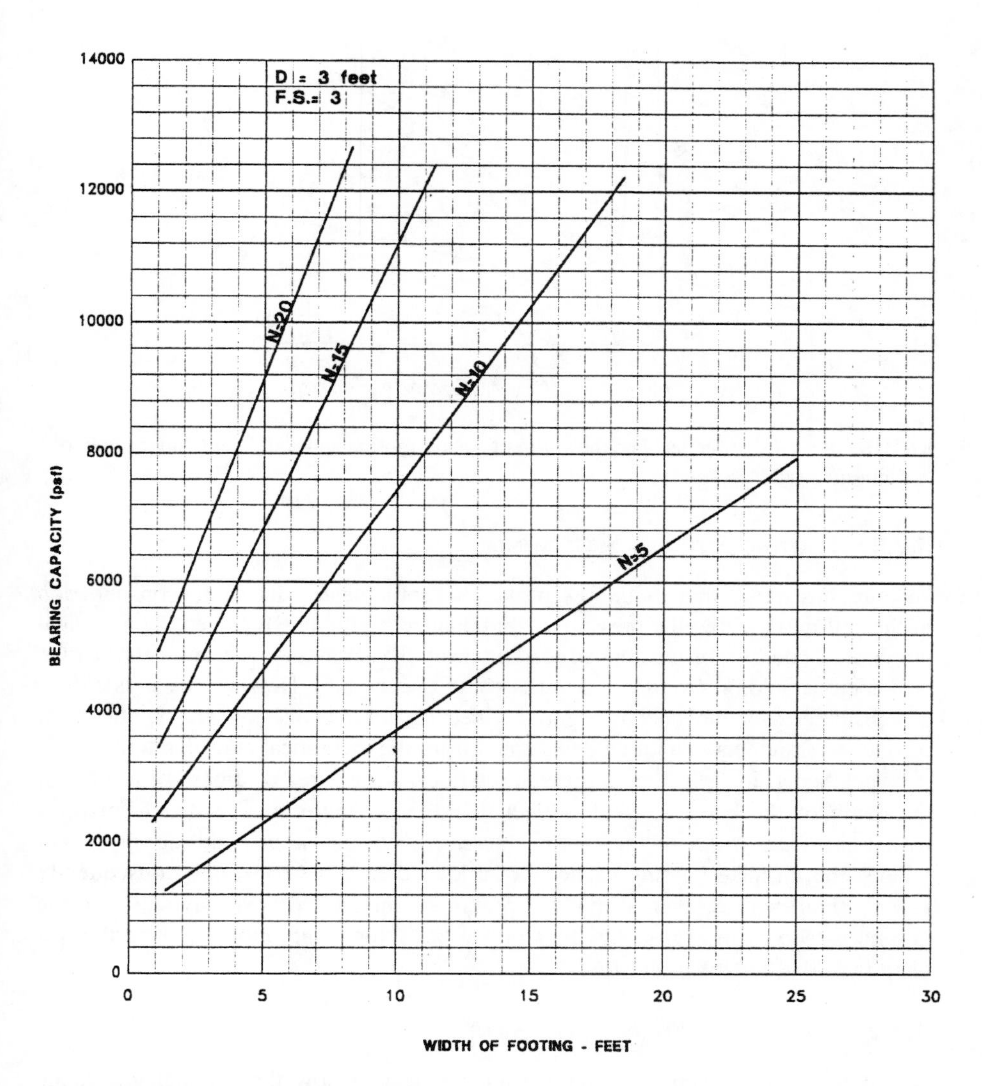

FIGURE 8.4 Width of footing versus bearing capacity for various penetration resistance values, as used by Meyerhof.

3. **Unit Weight** — Unit dry weight directly reflects the degree of compactness of the sand mass
4. **Water Table** — Since the submerged unit weight of sand is only about half that of moist or dry sand, the water table plays an important role in settlement.

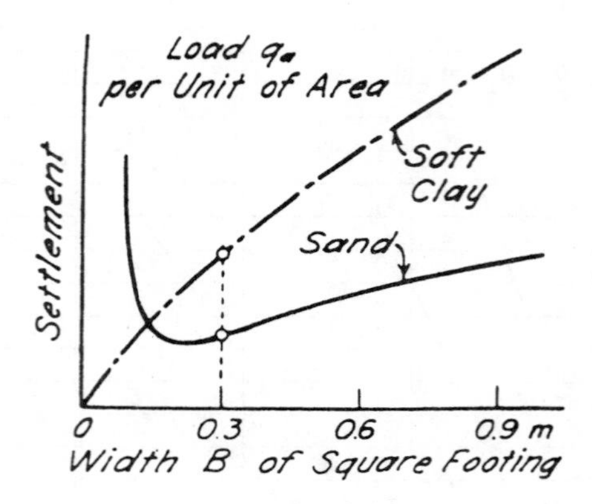

FIGURE 8.5 Relation between the width of square footing and settlement under same load per unit area (after Peck).

8.2.1 Footing Size and Settlement

Terzaghi has shown that about 80% of the total settlement is due to the consolidation of the soil mass within the pressure bulb, bounded by the line representing a vertical pressure of one fifth of the applied load intensity. By similarity, it is apparent that the settlement should be proportional to the width (Figure 8.5). However, as soils cannot be considered as homogeneous material, especially for the cohesionless soil, the effect of size on settlement cannot be determined from the theoretical considerations.

Peck stated in 1996 that at a given load per unit of area of a base of a footing, the depth of the body of sand subject to intense compression and deformations increases as the width of the footing increases. On the other hand, at very small widths, the ultimate bearing capacity of a loaded area is very small; consequently at very small widths, even at low soil pressures, the loaded area may sink into the ground as shown in Figure 8.5. This is one basis for commonly requiring the test plate in a load test be at least 4 ft^2.

8.2.2 Footing Depth and Settlement

For footings of a given size, the greater the depth below the original grade, the greater may be the allowable bearing pressure for a given settlement. It is not the depth that directly affects the results; the important factor is the ratio of the depth to the width (D/B), which is termed the depth factor. It was shown that for a depth equal to one half the width of the footings, the amount of settlement is only half of that in the surface loading condition.

Adjustment of the depth effect on the allowable bearing pressure on sand is seldom attempted. However, it is a general practice for geotechnical engineers to assign a higher bearing pressure for piers bottomed on sand than spread footings founded on the same material.

It must be emphasized that the depth effect is based on undisturbed, homogenous sands; disturbance to the soil during construction may affect the depth effect. Terzaghi

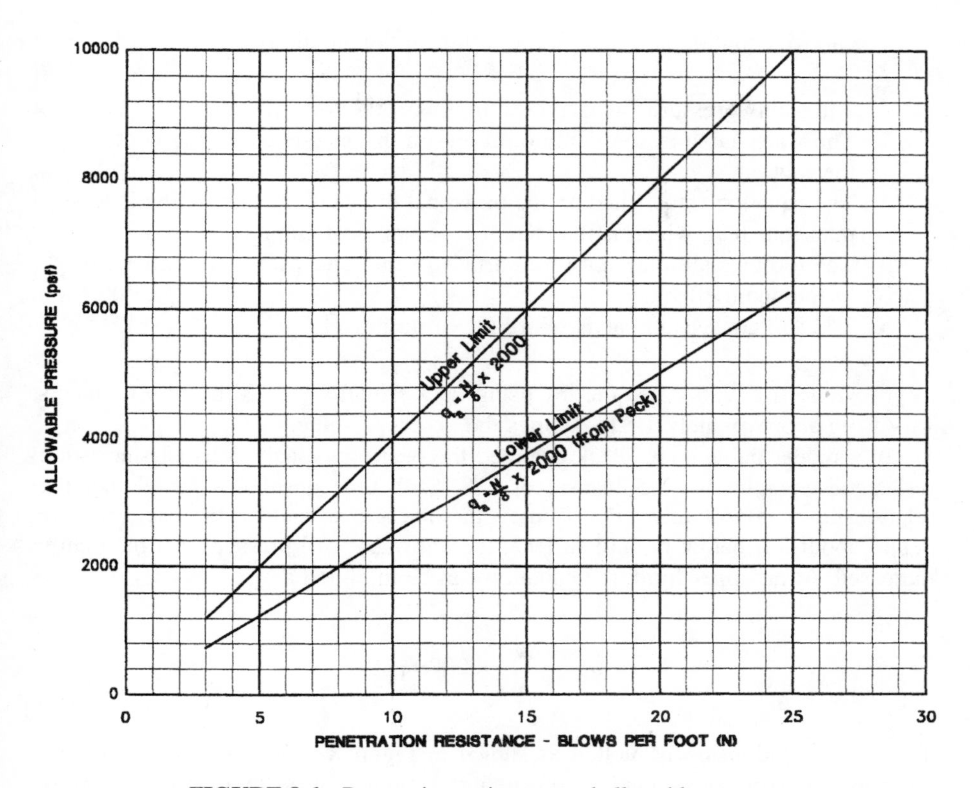

FIGURE 8.6 Penetration resistance and allowable pressure.

has called attention to the fact that the loosening of soil during the excavation of a deep shaft in sand may lead to settlements that are as large as those which would occur under the same loading at ground surface.

8.2.3 Penetration Resistance and Settlement

The simplest and easiest method in evaluating the bearing capacity of footings founded on granular soils is by correlating with the penetration resistance value (Figure 8.6). The standard penetration test, when performed on medium-grain gravel and sand, is reliable and easy to perform. As early as 1948, Terzaghi and Peck proposed the correlation of bearing capacity and penetration resistance with the following equation:

$$q_d = \frac{N}{8} \times 2000$$

where q_d = The allowable bearing capacity in pounds per square foot
N = Penetration resistance in blows per foot

The above equation is based on the following assumptions:

1. Allowable pressure is based on a footing settlement of 1 in.
2. The water table is at a depth of at least a distance equal to the width of the footing.
3. The equation established on the basis that the width of the footing is less than four feet, which is the size of footings most commonly used.
4. The footings are placed on the surface of the sand with no consideration of the depth effect.
5. The soil consists of sands with little or no gravel.

Meyerhof in 1965 stated that by using Peck's figure, the estimated settlements vary from approximately 1.5 to 3 times the observed value.

In a recent publication, Peck changed the bearing capacity versus penetration resistance equation by *N*/5 instead of *N*/8. Still, the consultants found that the relationship is conservative. For sands with some gravel and for the usual footing depth, about 3 ft below ground surface, the allowable soil pressure can be greatly increased. At an upper limit, the equation can be modified as:

$$q_d = \left(\frac{N}{5}\right) \times 2000 \text{ psf}$$

The above relationship is plotted as shown in Figure 8.6. In choosing the proper allowable pressure, geotechnical consultants should rely on their judgment more than charts and figures. The following should enter into consideration:

1. The percentage of gravel and cobbles in the deposit
2. The depth and width ratios of the footing
3. The amount of silt and clay in the deposit
4. The possibility of rise of the water table
5. The possibility of the development of a perched water condition

If all conditions are not favorable, use Peck's value. These conditions include the following: if the material does not contain a large amount of gravel; if the footings are placed near ground surface; and if the water table is near the base of the footings. On the other hand, if all conditions are favorable, and if neighboring structures have not suffered damage, then there is no reason why the upper limit as shown in Figure 8.6 cannot be economically used.

Using the above relationship, a convenient chart can be made for estimating the allowable soil pressure for various sizes of footings based on the penetration resistance data as shown in Figure 8.7. A similar chart can be prepared for the upper limit used by Peck. It is assumed that the footing width between 1 and 3 ft has the same allowable pressure. The chart is not applicable for a footing width of less than 1 ft.

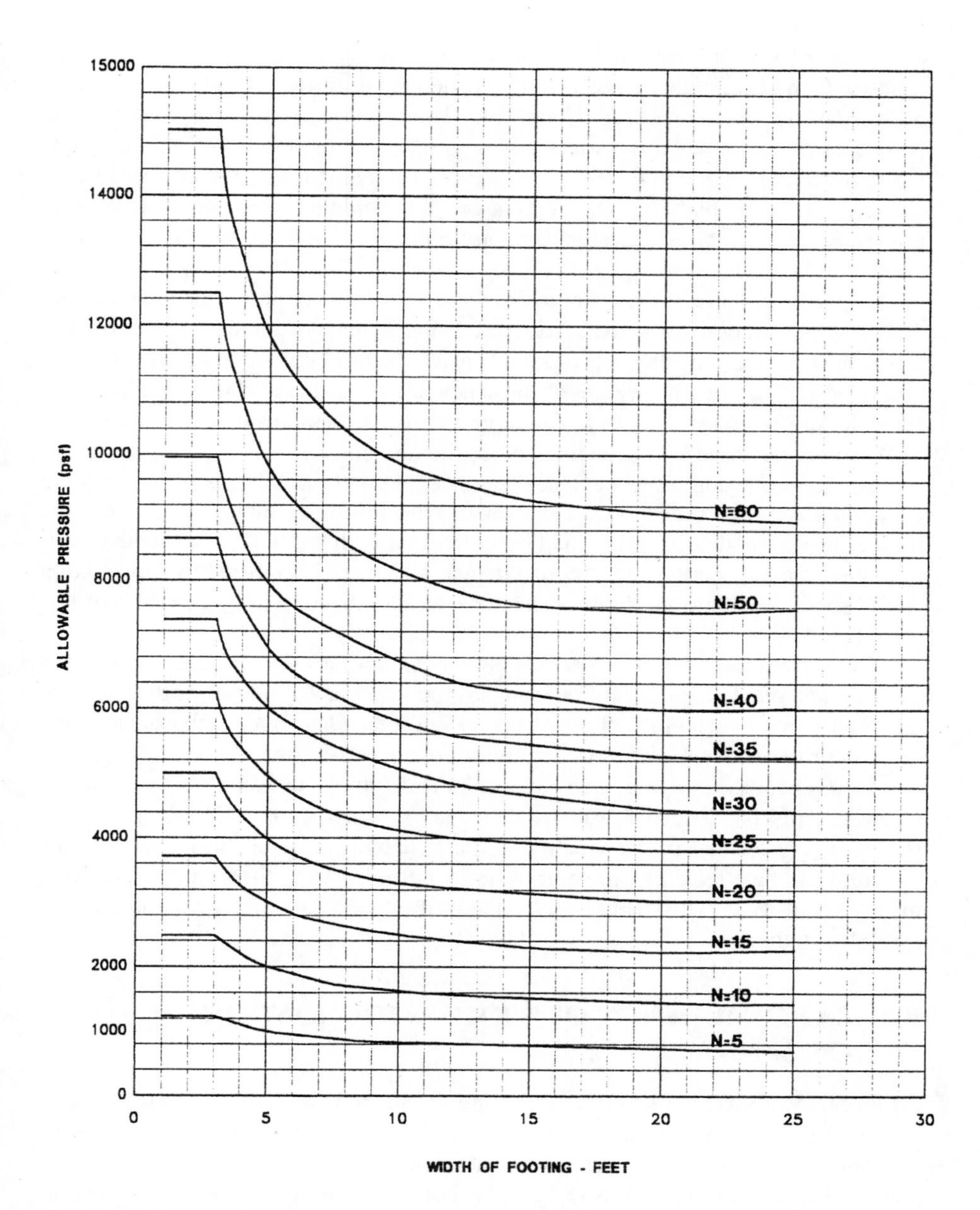

FIGURE 8.7 Relationship between *N* value and allowable pressure for maximum settlement of 1 in.

8.2.4 Water Table and Settlement

The position of the water table plays an important role in the determination of the stability of foundations on sandy soils. The water table elevation affects both the bearing capacity and the settlement of the foundation.

On the issue of correction for the water table, Peck commented in 1996 as follows: If the water table lies above the loaded depth of the footing, the confining pressure of the sand is reduced. Hence, the settlement correspondingly increases as compared to the values if the water tables were below the loaded depth. However, the reductions in confining pressure also cause the reduction in the standard penetration resistance value. The two effects largely compensate for each other. Therefore, the presence of a high water table can appropriately be ignored, and no water table correction is needed.

On the other hand, if the water tables were to rise into or above the loaded depth after the penetration tests were conducted, the actual settlement can be totally different. The confining pressure of sand beneath the footings and beside the footings is proportional to the unit weight of the sands. Hence if the sand mass has changed from the dry or moist state to a submerged state, the settlement of the footings is likely to be increased by as much as twice the amount.

It is important that the geotechnical engineer check the possibility of shear failure on narrow footings founded on loose sand, with the water table located near the footing level. Another aspect is the possibility that with the high water table condition, the ultimate bearing capacity of granular soils may be reduced by liquefaction due to shock or vibration. (Earthquake consideration is not within the realm of this book.)

The water table may not pose a problem at the time of investigation. But it is possible that due to local condition changes, the water table will rise or drop to such an amount that the stability of the structure will be endangered. A thorough investigation of the site is essential to determine such a possibility.

Throughout the site of a large structure, the depth of the water table may not be uniform. This is especially true when irrigation ditches or other water-carrying structures are located in the vicinity of the structures. Part of the footings in the structure can be affected by the high water condition while others can be free from the effects of water. Differential settlement of the footings is important and must be carefully studied.

8.3 RATIONAL DESIGN OF FOOTING FOUNDATION ON SAND

As discussed above, the stability of footing foundations on sand depends on many factors. Both from the standpoint of shear and settlement, these factors cannot be determined with certainty. Since almost all soils existing in nature are not homogenous, at a building site the soils vary in both vertical and horizontal directions. Only in rare cases can the theoretical analysis apply. The following details should be considered.

1. The basis of most theoretical analyses hinges on the value of the in situ penetration test value. Extensive research has been done in the past years to refine, correct, and correlate the field data. A field engineer admits that

the blow count data obtained greatly depends on the skill of the operator, the condition of the sampler, and the depth from which the tests are taken. Penetration resistance data cannot be treated as a mathematical function and applied to an equation as in treatment of steel or plastic. This is more evident when the blow count is below 4 or above 50.

2. A field engineer realizes that the penetration resistance value obtained can be totally different within a short distance of 10 ft at the same depth. For a given project, the number of borings and the frequency of the penetration tests taken are limited and the N value obtained at best can only give a general idea of the average subsoil condition.
3. The most important factor to be considered by the geotechnical engineers is the water table level. The groundwater level is important not only at the time of investigation but also in the future. Fortunately, the perched water conditions seldom exist in the granular soils.
4. Clean Sand (SW-SP) seldom exists. Most granular soils contain appreciable amounts of fines. A percentage of silt and clay, as much as 15%, is commonly encountered. In such cases, the settlement of the subsoil should be controlled by a consolidation test. Settlement estimates on silty sands should be treated in the same manner as the consolidation of clays.
5. Homogenous sand strata extending to a great depth seldom exist in nature. Pockets and layers of soft silt or clay can sometimes be present within the loaded depth of the footings. Such layers can easily be missed by the driller or noticed by the field engineer. Consolidation of such strata may control the settlement of the structure, and calculations of the settlement of sands become a minor item.
6. When a soft layer of clay is located below the sand strata, settlement of such a layer can control the behavior of the structure. Compressible clay layers may be located at a considerable distance below the footings.

8.3.1 Typical Design Example

From the detailed analysis of the design criteria to be considered in the design of foundation on sand, it is up to the geotechnical engineer to choose a rational and economical recommendation for the project. This can best be illustrated by the following typical case.

Project:	A single-story warehouse structure with a full live load.
Column load:	Varies from 50 to 200 kips.
Footing Depth:	3 ft below ground (below frost depth).
Field data:	Average penetration resistance $N = 10$ (within loaded depth).
Factor of safety:	Between two and three.
Allowable Maximum Settlement:	1 in. (differential settlement .75 in.).

From the previous discussions, namely:

Penetration resistance and allowable pressure (upper and lower limit). (Figure 8.6)
Width of footing versus bearing capacity (Meyerhof). (Figure 8.4)
Width of footing versus penetration resistance. (Figure 8.3)
Penetration resistance and allowable pressure (lower limit). (Figure 8.7)

By using $N = 10$ value, the relationship between width of footing versus allowable pressure can be established by using Figures 8.8 and 8.9. Also, for settlement consideration the relationship between width of footing and bearing capacity can be established by using Figures 8.6 and 8.7. This is shown in Figures 8.8 and 8.9.

In the upper limit with a footing width less than 2.5 ft and in the lower limit with a footing width less than 4.5 ft, the design is controlled by the shear consideration. The choice of the use of the upper or the lower limit for the final recommendation depends on the content of fines in the deposit, the water table condition, and the uniformity of the deposit. These are discussed as follows:

1. If the subsoil contains more than 15% of gravel and cobble and if the water table is located below the loaded depth of the footings with no possibility of rising, then the upper limits can be used with confidence.With a column load of 150 kips, the size of the footing should be on the order of 7 × 7 ft.
2. If the subsoil contains essentially sand with no appreciable amount of gravel and cobbles, and there is the possibility that the water table may rise to within the loaded depth of the footing, then the lower limit should be used. That is, the size of the footing should be on the order of 9 × 9 ft.
3. If other conditions are the same as above, but there is the possibility that the water table may rise to near the ground surface, then the footing size given above should be increased by as much as 50%. As an alternative, a reliable permanent dewatering system or drainage system can be installed to lower the water table. In such a case, the size of the footings can be reduced.

Irrespective of all the research and study done on the stability of the footing foundation on sand, the geotechnical engineer should use his or her past experience, keen observations, and common sense to achieve a logical and safe foundation system.

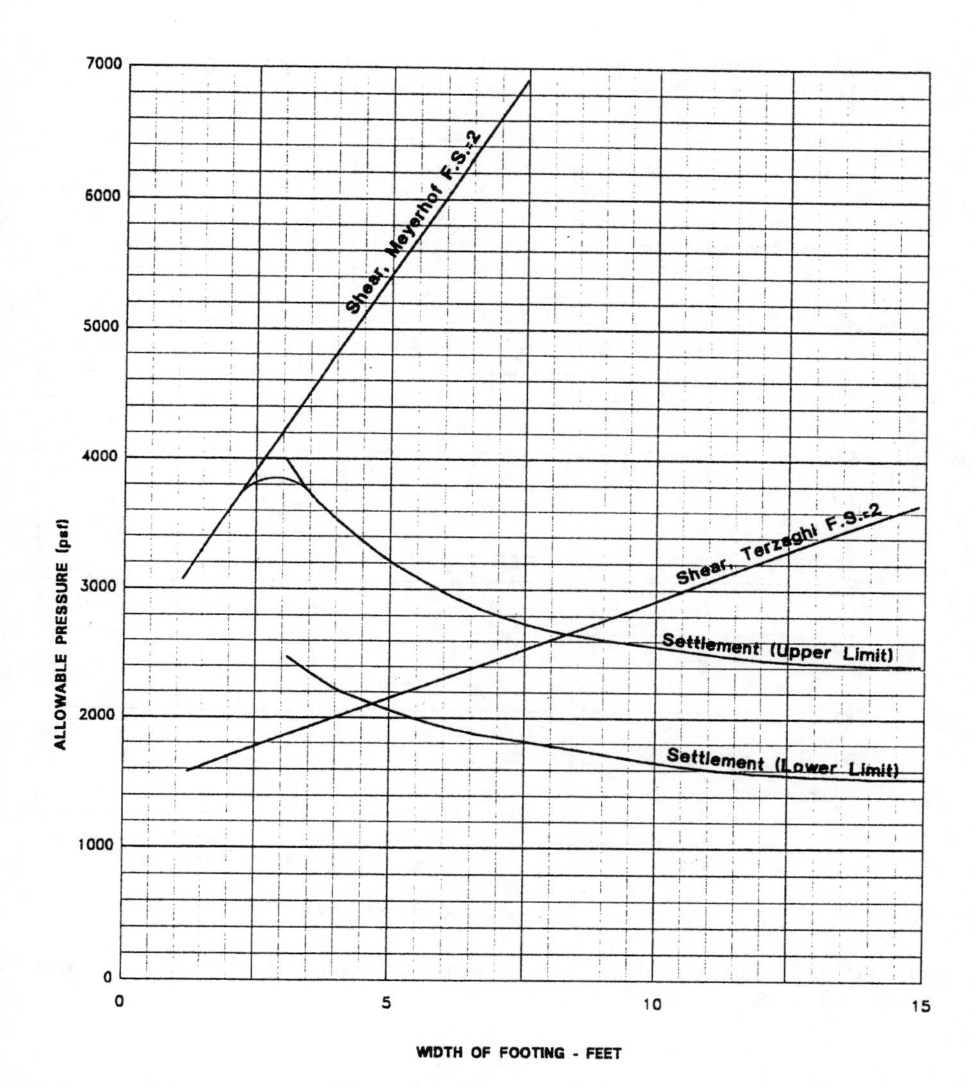

FIGURE 8.8 Width of footing versus allowable pressure. Upper and lower design limit for $N = 10$.

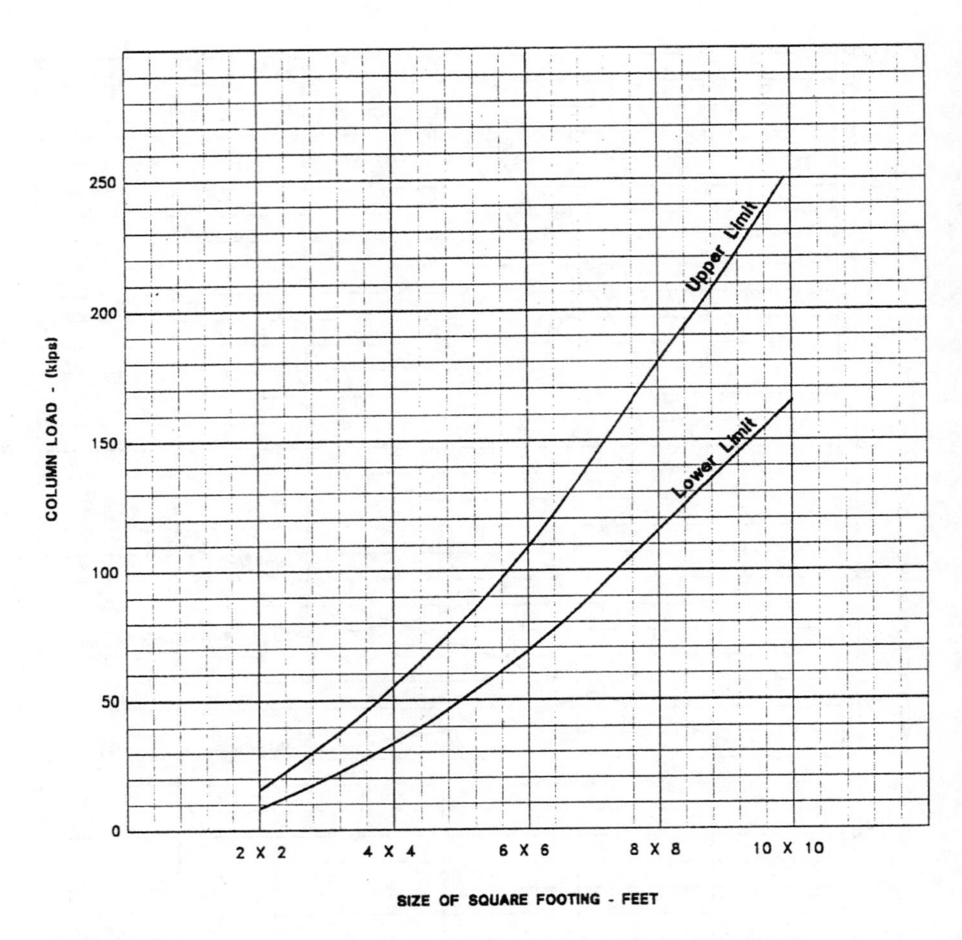

FIGURE 8.9 Allowable column load and appropriate footing size for upper and lower limits, with $N = 10$.

REFERENCES

G. G. Meyerhof, Shallow Foundations, ASCE Journal Soil Mechanics 91, No. SM 2, 1065, 1951.

G. G. Meyerhof, The Ultimate Bearing Capacity of Foundations, Geotec 2, 1951.

K. Terzaghi, R. Peck, and G. Mesri, *Soil Mechanics in Engineering Practice,* John Wiley-Interscience Publication, John Wiley & Sons, New York, 1996.

9 Footings on Fill

CONTENTS

Geologically speaking, a vast area of land is formed by transportation, either by the process of deposition by water, or being carried and then deposited by wind. Some such deposits are of recent origin and cause concern to geotechnical engineers.

Historically, engineers avoided the use of fill as much as possible. It is a tradition that all important structures should be founded on natural ground. Until recently, the use of fill was limited to the construction of earth dams, canal embankments, and road subgrades. Fill ground has been generally regarded as unsuitable for supporting even minor structures.

With the development of fill placement technique and with the introduction of compaction equipment such as the sheepfoot roller, the use of compacted fill to support structures has become an accepted practice.

9.1 FOUNDATIONS ON FILL

Today, geotechnical engineers realize that the study of fill placement is as important as the study of the behavior of sand or clay. Generally, the investigation of fill enters into the following categories:

Fill placement on undesirable natural soils
Removal of existing fill and replacement with compacted fill
Removal of existing fill and replacement with imported compacted fill
Removal of upper natural soil and replacement with compacted fill

With today's improved construction technologies, the use of fill has become a routine part of a geotechnical engineer's recommendation.

9.1.1 Fill on Soft Ground

When the natural soils have a very low bearing capacity and it is necessary to place a relatively heavy structure on it, it is possible to place a structural fill to distribute the imposed load. A thorough investigation is required to justify such an undertaking. Some of the factors to consider are outlined below:

1. To know the extent and thickness of the soft soil strata
2. The compressibility of the soft soil strata must be determined
3. Under certain loads, it is necessary to estimate the time required to complete the consolidation
4. The location of the water table is sometimes necessary to control the feasibility of such a project
5. The feasibility of the installation of a dewatering system
6. The availability of suitable fill material
7. The tolerable amount of settlement
8. The type of compacting equipment available

The most difficult problem confronting a geotechnical engineer is the erection of structures on very soft organic clay or silt. Such problems often rise during the construction of highways or railroads. Natural deposits of this type are common in regions formerly occupied by shallow lakes or lagoons. The deposits usually consist of peat moss or other types of marsh vegetation. Such soils may not be able to sustain the weight of a fill more than few feet in height. Fill on such foundations may continue to settle excessively for many years or decades.

During the construction of the Tibet Highway, a vast area of marshy ground was encountered (Figure 9.1). The deposit extended many square miles and was located at elevations above 18,000 ft. It was believed that the area constitutes part of the sources of the Yellow River. The deposit was so soft that it did not sustain even horseback riders. The subsoil contains about 50% silt and clay with a liquid limit of more than 100.

Since no granular soils were available, ditches were dug along both sides of the proposed roadway to a depth of about 10 ft to lower the water table. The excavated material was allowed to dry, then used as fill. The completed road was able to support truck traffic.

9.1.2 Removal and Replacement of Existing Fill

If the soft clay layer is thin and near the ground surface, it is more economical to remove the clay layer and replace it with compacted structural fill. Such an operation should be limited to the following conditions:

That the soft clay layer exists within about 10 ft below ground surface

That an excessive amount of settlement can take place if such a layer is not removed

FIGURE 9.1 Marshy ground near Tibet.

That under the soft layer, high bearing capacity soils exist, such as bedrock or stiff clays
That imported fill material is within economical reach

More critical than the soft clay layer, the existing fill may consist of by-products other than soils. These are trash materials such as building debris, concrete, bricks, ashes, cinders, and organic matter. Some of such materials can be used as fill, but the separation is difficult and the cost of such an operation can seldom be justified.

Coarse mine tailings are generally considered a good source of fill as long as they do not contain radioactive matter. The use of such materials as fly ash, chimney dust, etc., should be determined after laboratory testing.

When any of the above trashy material is suspected to be present, field engineers should be aware of the following:

1. The location of such material is erratic and cannot be determined. Drilling of test holes can sometimes miss the fill. In such cases, the geotechnical engineer faces an angry client. Laypersons may not understand bearing pressure, but they certainly can recognize trash material.
2. The trash fills can be very old and extend to a great depth. In one case, a historical building more than a century old in Denver, Colorado suddenly showed cracking on walls. Upon careful drilling, decomposed coffins were found at a depth more than 20 ft below the ground surface.
3. The extent of trash removal must not be limited to the proposed building line. It is sometimes necessary to remove as much as 15 ft outside the

FIGURE 9.2 Typical pattern of cracks due to deep-seated settlement.

building line in all directions. This is to prevent the trash material from affecting the stability of the foundation within the pressure bulb influence.

4. The presence of deep-seated trash fills can sometimes be revealed from the examination of the pattern of cracking of existing buildings. Patterns of cracking from differential settlement or heaving generally are in a diagonal pattern, while deep-seated fill settlements usually are in a horizontal pattern, as shown in Figure 9.2.

9.1.3 Recompaction of Natural Soft Soils

Such an operation is limited when the low-bearing soils are located within about 10 ft below the ground surface, and also when the bedrock is deep and pile or pier construction is costly. The use of such a system can best be illustrated by the following project:

Project	Ten-story hotel building.
Column load	Unknown at time of investigation.
Subsoil	Five to ten ft of loose silty or clayey sand. Average penetration resistance $N = 5$, with a few $N = 2$ underlain by 20 to 30 ft of medium dense clayey sands, with average penetration resistance $N = 15$. Claystone bedrock is at a depth of about 45 ft.
Water table	Stabilized at a depth of 58 ft.

The upper 10 ft of silty sands has a low-bearing capacity with a possibility of shear failure and cannot be used to support a high column load. The bedrock is deep and a pier foundation system is costly.

FIGURE 9.3 Stouffer's hotel, built with footings on compacted fill.

The most economical foundation system is to remove the upper 10 ft of low-bearing capacity sands and replace re-compacted, with the following requirements:

Excavation Remove at least 10 ft of the existing silty sands. Removal should extended at least 10 ft beyond the building line in every direction.

Compaction The removed soil should be re-compacted to at least 100% standard Proctor density at optimum moisture content.

Control Full-time inspection control is required by an experienced field engineer with frequent density tests performed.

Bearing capacity Footings placed on the controlled structural fill should be designed for a bearing pressure of 5000 pounds per square foot.

The completed structure is shown in Figure 9.3. The building is 20 years old with negligible settlement.

9.2 COMPACTION

The compaction of fill increases the bearing capacity of foundations constructed over them. Compaction also decreases the amount of undesirable settlement of structures and increases the stability of the slopes of the embankments.

9.2.1 Compaction Tests

The standard Proctor test as described in Chapter 5 is most commonly used. Highway and airport engineers choose to use the modified compaction test that offers high compaction effort. The Corps of Engineers adopted another set of standards for controlling its fill. It is possible to control the fill with one set of standards by varying the degree of compaction from, say, 90% to 100%. A great deal of confusion can be avoided by adopting a standard compaction test for all fill.

With the development of heavy rollers and their use in field compaction, the standard Proctor test was modified to better represent field conditions. The soil is compacted in five layers with a hammer that weighs 10 lbs. The drop of the hammer is 18 in. The number of hammer blows for each layer is kept at 25 as in the case of the standard Proctor test. The modified Proctor test is used most of the time by the Bureau of Reclamation as well as the U.S. Army Corps of Engineers.

Still, the procedures are not without loopholes. In one case, the soil compaction in the field was obviously inadequate, yet all field density tests indicated higher density than the Proctor maximum density requirement. After many months of litigation and investigation, it was found that the technician performing the Proctor density test placed the compaction cylinder on the open tailgate of his pickup truck instead of on a solid surface. Consequently, the maximum density was as much as 5 lbs lower than the actual value. This case indicates that no testing can replace experience and common sense.

A hand-dug hole is made in the fill and the removed soil collected for weight measurement. There are several methods for determining the volume of the hole. For many years, the consultants in the Rocky Mountain area chose the "sand replacement method," where the hole is filled with pre-calibrated Ottawa sand through a special sand cone device (ASTM D-1556). With this method of testing, the field engineer is able to gain a good feel of the condition of the fill while digging the test hole. However, the result of the test cannot be known until the laboratory completes the testing. The earth-moving contractor will not be able to know whether the compaction meets the required criteria for at least 24 h.

In recent years, the nuclear method for compaction control was widely used. Both the bulk density and the moisture content of the fill can be measured using controlled gamma radiation techniques. The apparatus generally consists of a small, shielded radiation source and a detector (Figure 9.4). The intensity of transmitted or back-scattered radiation varies with density and moisture content. Calibration charts are used which relate detected radiation intensity to values recorded from soil with known intensity. In 1980, the device was adopted by ASTM.

Consultants prefer to use such a device, as it reduces laborious hole digging. Contractors welcome such a device, as it provides immediate results. Some unreasonable test results have been attributed to a wrongly calibrated instrument. Field engineers should review the report carefully before submission. Field manual density determination methods are still used by consultants.

Proctor tests can be performed only with soils containing at least 6% of fines. For clean granular soil (SP-GP), relative density should be used. Since relative density and Proctor density originated from totally different concepts, no correlation should be established between them. When preparing specifications for a certain project, some tend to copy the old specification without giving consideration to the difference between standard Proctor, modified Proctor, or relative density, which results in specifying an unnecessary degree of compaction and boosting the cost of the project. In some unfortunate cases, the problem has to be decided in court.

FIGURE 9.4 Nuclear density test.

Specifications generally call for compacting the soil to an optimum moisture content. Since it is not possible to obtain an exact optimum moisture content for every test, it is necessary to specify the amount of deviation permitted. Field engineers and contractors usually are very strict in meeting the compaction requirements and pay little attention to the moisture content. For a well-compacted fill, the moisture requirement is equally important as density.

9.2.2 Compaction Equipment

The most commonly used equipment for the compaction of fill is:

Smooth-wheel roller
Pneumatic rubber-tired roller
Sheepsfoot roller
Vibratory roller
Jumping tamper

In many projects, the contractor claims that the specified degree of compaction is impossible to obtain. In fact, it is only the wrong equipment that is responsible for the lack of density. Some earth movers claim that with their equipment they are able to compact the soil with lifts as thick as 2 ft. Field engineers should never believe such claims, since no known equipment is able to compact soils to the specified density with more than a 12-in. lift.

Smooth-wheel rollers are suitable for proof rolling subgrades and for finishing operations of fill with sandy and clayey soil. In developing countries, rollers are carved from solid rock and pulled by a team of laborers. A satisfactory degree of compaction can be achieved when used in thin layers.

Pneumatic rubber-tired rollers are heavily loaded wagons with several rows of closely spaced tires. The load is provided by a ballast box, which is filled with earth or water. The widely used 50-ton roller applies its 25,000-lb wheel load at 100 psi to an equivalent circle having an 18-in. diameter. The compaction is achieved by the combination of pressure and kneading action.

Sheepsfoot rollers are drums with a large number of projections. They are most effective in compacting clay soils. The medium rollers are capable of producing densities greater than the standard Proctor maximum in lifts 6 to 12 in. thick and at a moisture slightly below the optimum at six to eight passes over the surface. Sheepsfoot rollers are the most commonly used equipment of the earth contractor.

Vibratory rollers are very efficient in compacting granular soils. Vibrators can be attached to smooth-wheel, rubber-tired, or sheepsfoot rollers to provide vibratory effects to the fill. Hand-held vibratory plates can be used for effective compaction of sandy soils over a limited area, such as in backfill along basement walls.

Jumping tampers are actuated by gasoline-driven pistons that kick them into the air to drop back on the soil. A typical jumping tamper weighs 200 lb, jumps as high as 18 in., and delivers a blow capable of compacting soils in layers 6 to 12 in. thick to the standard Proctor maximum density at optimum moisture content. Such hand-operated tampers are frequently used in compacting backfill or fill placed in tight areas where large equipment cannot reach.

9.2.3 Compaction Control

Various field compaction control methods are described in Chapter 5. It should be emphasized here that all methods are subject to certain limitations. The presence of large cobbles can throw the results off, unless a careful rock correction factor is applied. For certain projects, the Corps of Engineers specifies the use of an 8-ft diameter ring, removing all soil within the ring and filling the excavated hole with water. Such tests are accurate and can represent the entire project, but they are costly and time-consuming.

Today, almost all compaction tests are performed with the use of a nuclear density device. Such tests enable the engineer to reject or approve the compaction immediately. However, since all fill materials are erratic in composition, test results must depend on the frequency of the Proctor density test.

The control of fill placement requires much more than density tests. An experienced field engineer should observe the fill operation to determine its adequacy. He or she should observe the following:

1. The type of compaction equipment. If a sheepsfoot roller is used, check the size of the roller and the ballast weight.
2. Check whether the equipment is suitable for compacting the type of soil. Vibratory rollers should not be used to compact heavy clay.
3. The number of units of compacting equipment should be proportional to the yardage of earth to be removed.

4. A sufficient number of passes (generally six to eight) should be made over the same area to ensure proper compaction. This is more important than a density test.
5. A sufficient number of water trucks should be available to ensure the correct moisture content. In 9 out of 10 small projects, the moisture content of the fill is below the specified amount.
6. Generally, the thickness of each lift should not exceed 12 in. It is a false claim that the earth mover is able to compact the fill with several feet of lift thickness.
7. If the fill is prewetted at the stockpile, be sure to check the moisture before spreading.
8. Find out the construction experience of the superintendent.

The above checklist is more important than a few density tests performed over a large fill project. After all, the general stability of the fill is determined by the overall performance.

9.2.4 Degree of Compaction

The degree of compaction of a structural fill depends on the function of the project and the tolerable settlement. Generally, for fill supporting footings with 95% compaction, the allowable soil pressure should be on the order of 3000 to 4000 psf. Compaction of 100% will be required for allowable pressure in excess of 4000 psf. The thickness of the fill and the characteristics of the underlying natural soil also have a strong bearing on the allowable soil pressure of fill. Careful engineering should be exercised if the soil beneath the fill consists of soft clays. Excessive settlement can take place due to the weight of the fill. In no case should the thickness of the fill beneath the footings be less than 24 in. In order to control differential settlement, all footings should be placed on uniform fill. The practice of placing a portion of the footings on fill and another portion on natural soil should be avoided if at all possible.

The degree of compaction for fill supporting floor slabs is generally less critical than for footings. Exceptions are heavy-loaded floors supporting vibratory machinery and floors supporting vehicle loads. Compaction of 90 to 95% should be achieved in each case.

The evaluation of rock and boulder for use in fills is more difficult. Fill consisting of high percentages of large rocks is not recommended for use on structural fill projects, since uniform compaction is difficult to achieve.

Cohesive soils often lose a portion of their shear strength upon disturbance. The amount of strength loss due to disturbance is expressed in terms of "sensitivity." An undisturbed sample and a remolded sample of the soil with the same moisture content and density are subjected to unconfined compressive strength tests. The ratio between the undisturbed strength and the remolded strength is the sensitivity of the soil. Clays with sensitivities of four to eight are commonly seen.

Therefore, it is not fair to compare the density of compacted fill with that of natural soil of the same clay. Clay, after recompaction, although having greater

To Meadows Metropolitan Districts 1-8
6060 South Willow Drive, Suite 2101
Englewood, Colorado 80111

Attention: Mr. Glen Smith

Job No. 1 405 86 Date 10/7/86
Daily Report No. 75 Sheet 1 Of 1
Subject: Fill Observation and Testing
Meadows Boulevard, Castle Rock, Colorado

TEST NO.	LOCATION	DEPTH OR ELEVATION (FEET)	LABORATORY MAXIMUM DRY DENSITY (pcf)	LABORATORY OPTIMUM MOISTURE CONTENT (%)	FIELD DRY DENSITY (pcf)	FIELD MOISTURE CONTENT (%)	PERCENT COMPACTION	SOIL TYPE
455	15' North of South End East Side	11'BTCB	104.5	12.5	102.7	13.9	98	Very Sandy Clay
456	40' North of South End East Side	9'BTCB	104.5	12.5	101.0	13.7	97	"
457	Centerline East Side	9'BTCB	104.5	12.5	101.0	13.0	97	"
458	Centerline East Side	8'BTCB	104.5	12.5	98.4	12.7	94	"
458A	Centerline East Side	8'BTCB	104.5	12.5	100.6	12.4	96	"

SPECIFICATION COMPACTION & MATERIAL 95% standard Proctor density

TYPE AND NUMBER OF EARTH MOVING UNITS (1) Dozer (2) Scrapers
TYPE AND NUMBER OF COMPACTION UNITS (1) 815 Sheepsfoot
NUMBER OF PASSES As Req'd THICKNESS OF LIFT Unknown
METHOD OF ADDING MOISTURE Water Truck

This report presents opinions formed as a result of our observation of fill placement. We have relied on the contractor to continue applying the recommended compactive effort and moisture to the fill during times when our observer is not observing operations. Tests are made of the fill only as believed necessary to calibrate our observer's judgement. Test data are not the sole basis for opinions on whether the fill meets specifications.

COPIES Town of Castle Rock; TST; HNIB

CAF-1R (6 85)

☒ FILL TESTED MEETS SPECIFICATIONS.
☐ FILL TESTED DOES NOT MEET SPECIFICATIONS AS INDICATED BY TEST NO.(S) AND SHOULD BE REMOVED OR REWORKED.
☒ CONTRACTOR ADVISED
☐ FULL TIME OBSERVATION ☒ PART TIME OBSERVATION

PROGRESS REPORT Hudick Construction continued to place and compact the backfill material for pedestrian underpass number 3.
BTCB=Below Top of Box

Tim Burkhard — FIELD OBSERVER
David M. Dix/djb — APPROVED BY

FIGURE 9.5 Typical compaction test data.

density, may not have the same strength as the undisturbed soil. The density of compacted soil cannot be compared with the density of natural soils. Earth contractors usually claim that the fill has density better than the natural soil. It is comparing apples with oranges. Taking density tests on natural soil in the course of fill control is not warranted.

A specification contract between the owner and the contractor is prepared to ensure that the required field density is achieved. It is generally specified that one density test be performed for every 2400 cubic yards of fill placed.

The consultant should avoid issuing statements such as "The fill was placed in accordance with the specification," or "The degree of compaction meets the specified requirement."

A typical fill control report form is shown in Figure 9.5.

9.2.5 Fill Used as Form Work

An unusually shaped roof consisting of several semi-circles was used for a religious center. The architect thought it was possible to use compacted fill as form work instead of costly timber frames. The fill was compacted to only 85% Proctor density, molded to the desired shape, and covered with plastic. The concrete roof was then poured immediately before the fill settled. The completed structure is shown in Figure 9.6.

FIGURE 9.6 Compacted fill used as form work.

REFERENCES

B.M. Das, *Principles of Geotechnical Engineering,* PWS Publishing, Boston, 1993.

S.J. Greenfield and C.K. Shen, *Foundations in Problem Soils,* Prentice-Hall, Englewood Cliffs, NJ, 1992.

G.B. Sowers and G.S. Sowers, *Introductory Soil Mechanics and Foundations,* Collier-Macmillan, London, 1970.

10 Pier Foundations

CONTENTS

A drilled pier, or a drilled caisson, is a cylindrical column that has essentially the same function as piles. The drilled pier foundation is used to transfer the structural load from the upper unstable soils to the lower firm stratum. The use of piers covers a wide range of possibilities, including these:

1. Drilling into bedrock for supporting a high column load
2. Transferring a bridge load to a level below that of the deepest scour
3. Friction piers bottomed on stiff clays for supporting a light structure
4. Belled piers bottomed on granular soils for supporting a medium column load

5. Long, small-diameter piers drilled into a zone unaffected by moisture change in swelling soil areas
6. For structures transmitting horizontal or inclined loading

Essentially, piers and piles serve the same purpose. The distinction is based on the method of installation. A pile is installed by driving and a pier by auger drilling. A single pier is generally used to support the column load while a pile group must be used.

10.1 LOAD BEARING CAPACITY

The bearing capacity of drilled piers is determined by its structural strength and the supporting strength of the soil. In almost all cases, the latter criteria takes control. The load carried by a pier is ultimately borne by either friction or end bearing. The load is transmitted to the soil surrounding the pier by friction between the sides of the pier and the soil, and/or the load transmitted to the soil below the bottom of the pier.

$$Q_{ultimate} = Q_{friction} + Q_{end}$$

where $Q_{ultimate}$ = Ultimate bearing capacity of a single pier
$Q_{friction}$ = Bearing capacity furnished by friction or adhesion between sides of the pier and the soil
Q_{end} = Bearing capacity furnished by the soil just beneath the end of the pier.

In many localities where bedrock can be reached, the piers are bottomed or drilled into the sound material. The allowable bearing pressure is specified by the building code. The code recommendations differ considerably, but generally they are conservative, and the type of bedrock is not clearly defined.

10.1.1 CONVENTIONAL APPROACH

Piers bearing on bedrock have been designed using mostly empirical considerations derived from experience, limited load test data, and the behavior of existing structures. Because of the erratic characteristics of bedrock, the ultimate load capacity of the Denver blue shale has never been determined. Nevertheless, piers supporting a column load in excess of 1000 kips are commonly designed and constructed in the Rocky Mountain region.

On empirical design of piers, Woodward, Gardener, and Greer stated, “Many piers, particularly where rock bearing is used, have been designed on strictly empirical considerations which are derived from regional experience.” They further stated that, “Where subsurface conditions are well established and are relatively uniform, and the performance of past construction well documented, the design by experience approach is usually found satisfactory.”

High-capacity piers in the Rocky Mountain area are primarily used to support the high-rise structures in the downtown area. The piers are founded in the claystone-siltstone bedrock, which has somewhat uniform characteristics.

The methods for determining the bearing capacity are as follows:

1. Penetration resistance of bedrock
2. Unconfined compressive strength
3. Consolidation tests on bedrock sample under high load
4. Actual record of settlement of existing building founded on similar bedrock
5. Load test on bedrock
6. Menard pressuremeter test on bedrock in the test hole

Unfortunately, all of the above approaches have their limitations. Penetration resistance on hard bedrock involves a blow count in excess of 100. When a soft seam or very hard lens is encountered, the actual penetration resistance of the stratum cannot be accurately determined. Therefore, to obtain an accurate determination of penetration resistance, many tests, possibly more than 50, should be conducted. After the representative penetration resistance value has been determined, the geotechnical engineer should convert the blow count to an allowable bearing capacity. One widely used method in the Rocky Mountain area is:

$$q_a = \frac{N}{2} \text{ (in ksf)}$$

where q_a = allowable bearing pressure
N = blow count

For instance, if the representative blow count is 40, the allowable bearing pressure selected will be 20,000 psf. This approach is quite conservative. Extensive laboratory testing indicates that a more realistic value for q_a should be in the range of N to 0.75 N. That is, with $N = 40$, the allowable bearing capacity should be nearer 30,000 to 40,000 psf.

The allowable bearing capacity derived by penetration resistance data should be checked by both the unconfined compressive strength test and the consolidation test. Due to sample disturbance, an unconfined compressive strength test at best can represent only the lower limit of the bearing capacity of bedrock. Unconfined compressive strength performed on drive samples can only be used for comparative purposes. Samples obtained from core drilling are far more reliable, but the presence of slickensides and other effects of disturbance limit the validity of such tests.

A high-capacity consolidation test on a reasonably good sample can sometimes be used to determine the amount of settlement, as shown in Figure 10.1. The actual settlement value taken as a percentage of laboratory consolidation value should be left to the judgment of the geotechnical engineer. In most cases, the actual pier settlement is only a fraction of the laboratory consolidation test value.

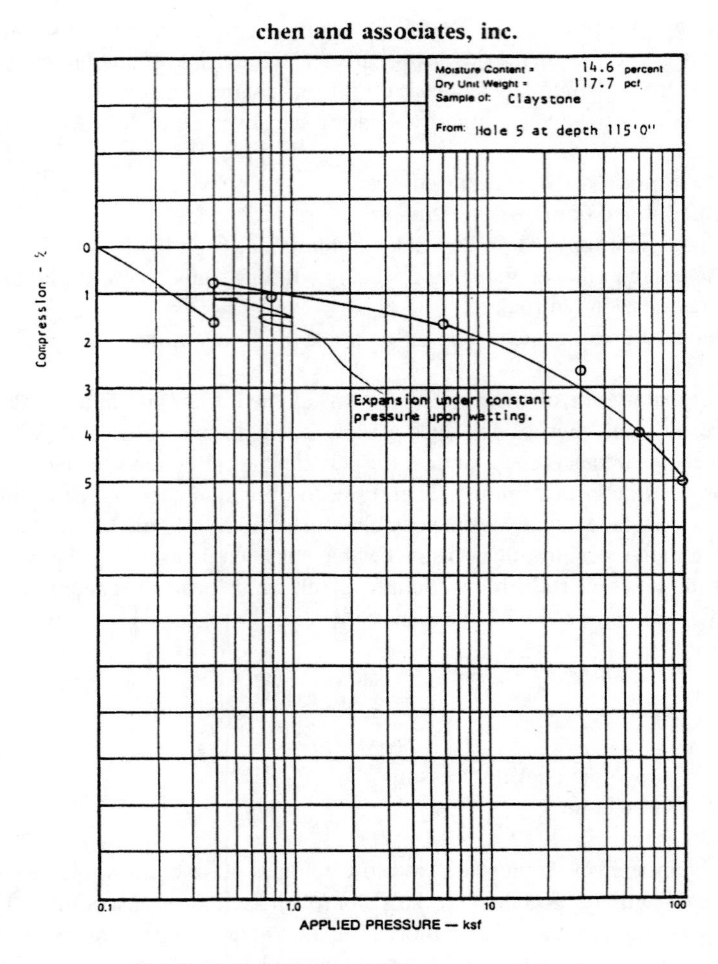

FIGURE 10.1 Consolidation test on bedrock.

10.1.2 Pressuremeter Approach

The most direct method for determining bearing capacity is the insertion of a Menard pressuremeter into the bottom of drill hole to evaluate the actual bearing capacity of the bedrock. The pressuremeter device is briefly described in Chapter 3.

The pressure increment and the volume change in the bore hole are plotted to develop the curve, as shown in Figure 10.2.

The initial curved portion occurs as the probe pushes the sides of the drill hole to its original shape. At this point, the at-rest condition of the ground is achieved and the relation between pressure and the volume change becomes a straight line. The slope of this straight line is the Menard modulus of deformation. The end coordinate of the straight line is known as the "creep pressure" since it marks the beginning of the plastic deformation phase of the test.

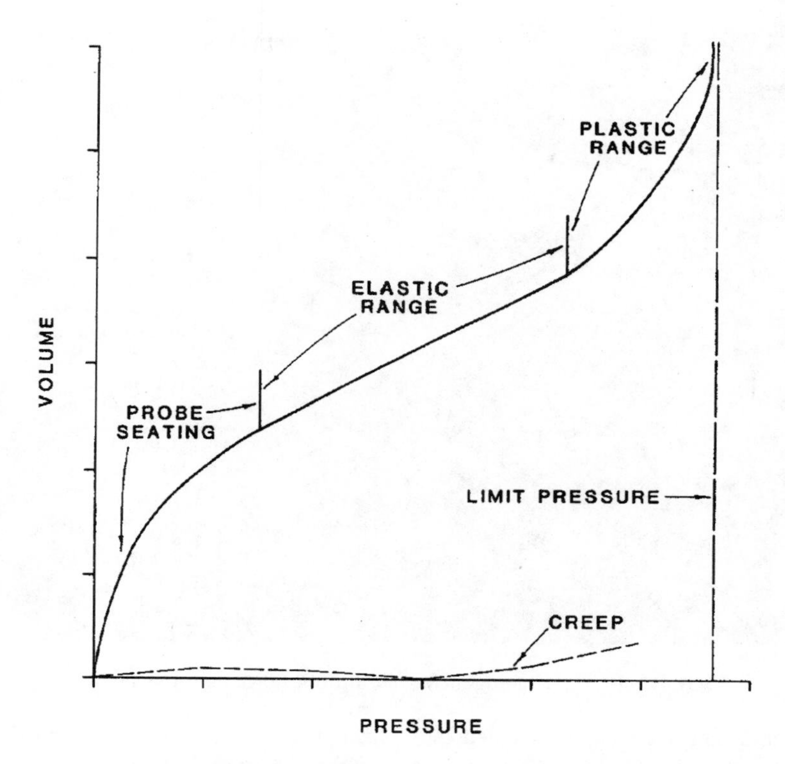

FIGURE 10.2 Typical pressuremeter test result.

Pressuremeter tests were conducted by Chen and Associates at Denver, Colorado for high-rise building foundations founded on large-diameter piers in claystone bedrock (Figure 10.3). The modulus of deformation values varied from 10,000 psf to 90,000 psf with a limit pressure varying from 72,000 psf to 344,000 psf.

High-capacity drilled pier shafts in downtown Denver vary in diameter from 4 to 9 ft with penetration into the claystone bedrock up to 30 or more feet. Loads on the piers vary to more than 6000 kips. Generally, the piers are designed for allowable end-bearing pressures of 40,000 to 100,000 psf. Such values exceeded the highest value of 80,000 psf recommended in the building code for shale.

The performance of high-capacity piers has been excellent. Settlement of the piers is seldom monitored; however, to date, the magnitude of pier settlement has not resulted in building distress of any structure known.

10.2 SKIN FRICTION CAPACITY

The load-carrying capacity of a pier depends not only upon its end-bearing value but also to a great extent on the skin friction or the side shear value between the concrete and its surrounding soils. The skin friction value is unfortunately difficult to determine. It is generally recognized that the skin friction value between cohesive

FIGURE 10.3 Large-diameter pier drilling for a high-rise building.

soils and pier shaft cannot exceed the cohesion of the soil. In stiff or hard clays, however, the bond between the concrete and the soil may be less than cohesion. It is generally recognized that the maximum design value for skin friction of all cohesive soils should be limited to about 1000 (or up to 1500) psf.

10.2.1 Conventional Approach

From the design standpoint, it appears the most significant variable affecting the shaft resistance or the skin friction is the shear strength of the supporting bedrock. Various approaches have been suggested in evaluating the skin friction value. The methods can best be illustrated by using the following data:

End-bearing	claystone bedrock
Penetration resistance	$N = 60$ blows per foot
Unconfined compressive strength	$q_u = 30$ ksf
End-bearing capacity	$q_a = 30$ ksf
Shear strength	$f = 15$ ksf

1. Choosing the establishment of skin friction by assuming that skin friction value in bedrock is "one tenth of the end-bearing value of bedrock," then for an end-bearing value of 30 ksf, the skin friction value is 3 ksf. It is generally thought that the skin friction value may not develop fully along the surface of the shaft, that a lubricated surface may exist between the soil and shaft.

2. David McCarthy estimated the skin friction values for drilled pier in clay as shown in Table 10.1.

TABLE 10.1
Skin Friction Values for Drilled Piers

Foundation Type and Drilling Method	Skin Friction (f)	Upper Limit on (f) Value (ksf)
Straight shaft, drilled dry	0.5 c	1.8
Straight shaft, drilled with slurry	0.3 c	0.8
Belled, drilled dry	0.3 c	0.8
Belled, drilled with slurry	0.15 c	0.5

Where c is soil cohesion = $q_u/2$

For an end-bearing value of 30 ksf, the unconfined compressive strength value of 30 ksf, and cohesion value 15 ksf, using the above table for 0.5 c the skin friction value should be $0.5 \times 15 = 7.5$ ksf; that is, more than the upper limit of 1.8 ksf. Therefore, the value 1.8 ksf should be assigned as the skin friction value.

3. Horvath and Kenney proposed a correlation between allowable side shear (skin friction) and shear strength, as shown in Figure 10.4.

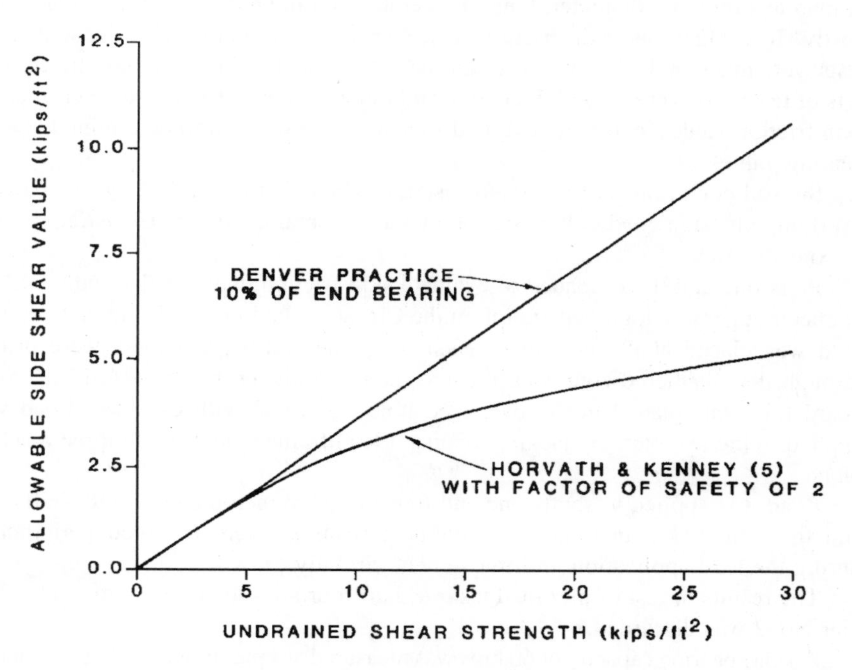

FIGURE 10.4 Shear strength vs. allowable side shear.

Also plotted in Figure 10.4 is the Denver practice of utilizing 10% of the shear strength for the allowable skin friction.

4. It is seen from the above illustration that the skin friction value established by various methods differs greatly.

Denver practice by 10% of the end bearing	3.0 ksf
Horvath and Kenney method	2.6 ksf
David McCarthy method	1.8 ksf
Denver Practice by 10% of the shear strength	5.0 ksf

5. In designing the pier capacity, the skin friction developed in the portion of pier embedded in the overburden soil is usually ignored. Such values are considered to be the factor of safety in the pier system. Actually, for long piers, the skin friction developed in the overburden soils can be considerable. It is important that an adequate factor of safety be incorporated to allow for unforeseen construction contingencies.
6. Despite the satisfactory performance of large-diameter, high-capacity drilled piers, it appears that the design and mechanism of load transfer are not completely understood. Sufficient differences exist between existing design procedures to warrant significant additional investigation into this problem.

10.2.2 Pier Load Test

A load test on large-diameter, long piers embedded in bedrock is difficult and very costly. To evaluate the skin friction with respect to the end-bearing capacity, load tests were made on low-capacity, small-diameter piers by Chen and Associates. Two sets of test piers were drilled. Piers one and three were constructed to determine the skin friction values, and piers two and four were constructed to determine the end-bearing values.

Subsoil conditions at the site consisted of about 2 ft of sandy clay overburden, overlying claystone bedrock. Core samples were obtained on the Claystone for the triaxial shear test.

Piers one and three, loaded to evaluate skin friction, were drilled with a 12-in. diameter auger to a total penetration in the Claystone bedrock of 5.5 ft. A cardboard void was placed at the bottom of each pier. The end-bearing piers were drilled through the weathered zone and 5 ft into the unweathered bedrock. A 12-in. cardboard tube was placed in the oversized hole and filled with concrete. Load was applied to the test piers by jacking against a load frame connected to three reaction piers.

Load was applied to shafts one and four in 10 kip increments and to shafts two and three in 20 kip increments. Settlement readings were recorded periodically during the load application until movement virtually ceased.

The results of test of pier load test on shaft four are shown in Figure 10.5. (Test pier No. 2 was discarded.)

An end-bearing capacity of 60 kips was measured at a pier movement of 1.2 inches, 10% of shaft diameter. The calculated in situ shear strength was approximately 8.5 ksf.

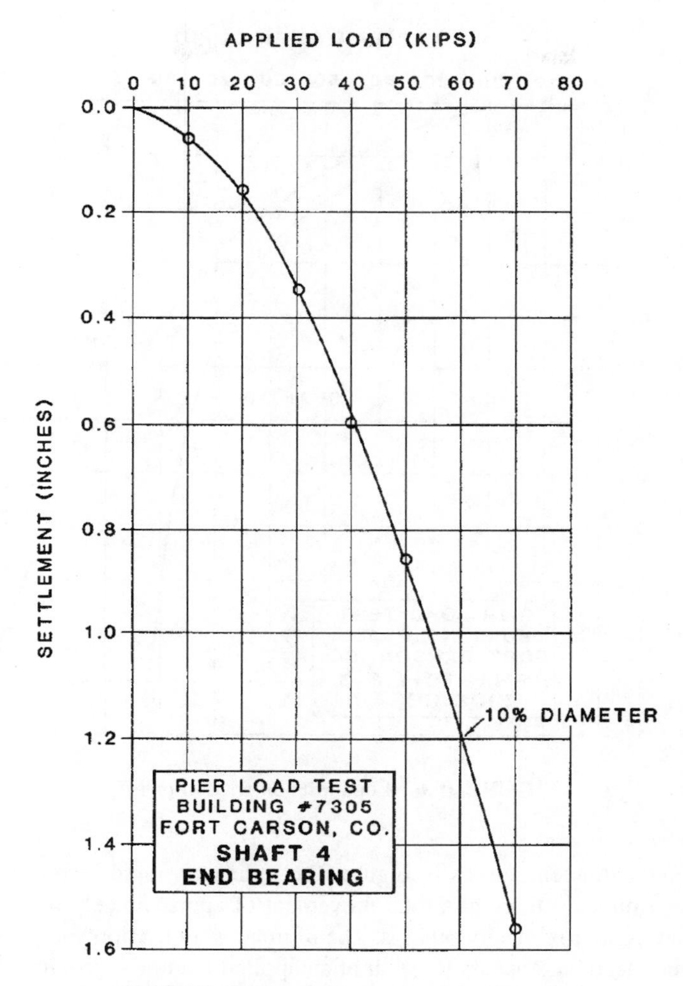

FIGURE 10.5 Load test on end-bearing pier.

with a factor of safety equal to one and bearing capacity factor 9 (i.e., 60/9 × 0.785) (Figure 10.6).

The following data are used for the evaluation of the load test results:

α = Ratio of the peak mobilized shear stress to the shear strength, averaged over the area of the shaft
A = Area of pier shaft embedded in shale, 3.14 × 5.0 × 1 = 15.7 ft^2
D = Diameter of pier shaft (ft)
N_c = Bearing capacity factor; use 9 for piers
s = Average triaxial shear strength = 11.3 ksf
F.S. = Factor safety. Use one.

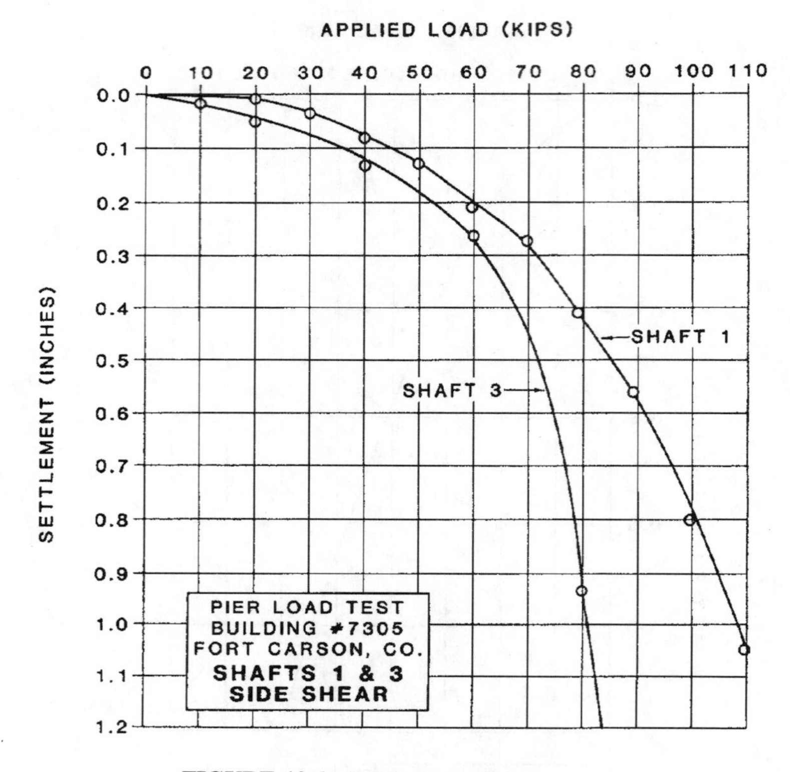

FIGURE 10.6 Load tests on friction piers.

The load settlement curves in Figure 10.6 indicate a rapid increase in the rate of settlement on shaft three at a total movement of approximately 0.25 in. with an applied load of approximately 60 kips. The ultimate skin friction value on shaft one is less pronounced but appears to occur at an applied load of approximately 70 kips at 0.28 in. Back calculation yields an approximate skin friction value of 4.45 ksf (70/15.7) and 3.82 ksf (60/15.7) on test piers one and three, respectively.

Comparing the ultimate skin friction values to the measured in situ shear strength obtained from a load test on shaft four, α values of 0.52 (4.45/8.5) and 0.44 (3.8/8.5) are obtained for shafts 1 and 3, respectively. Utilizing the average in situ strength from laboratory test results, 11.3 ksf, α values of 0.39 and 0.33 are obtained for shafts one and three.

The α values calculated from the load test results appear lower than previously reported for shale materials in this range of shear strength. Goeke and Hustad report an α value of 0.8 in a Cretaceous shale with a shear strength of 10.8 ksf.

Unfortunately, load tests on piers are seldom performed. Rational design of small- or high-capacity piers depends on the experience of geotechnical engineers and the performance of existing structures.

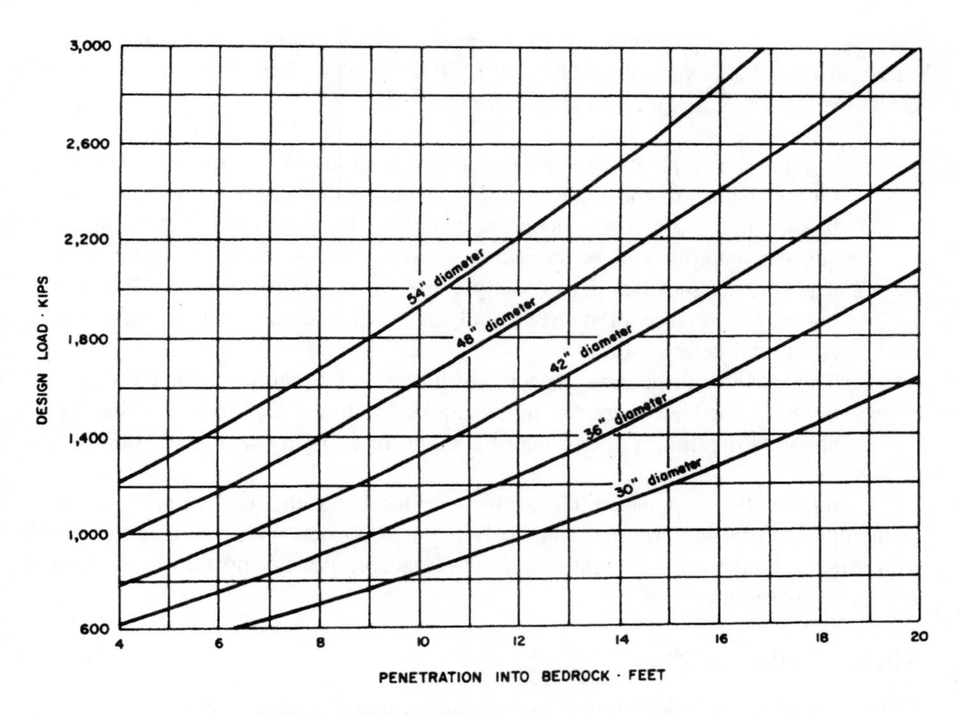

FIGURE 10.7 Design load versus depth into bedrock.

10.3 RATIONAL PIER DESIGN

With the above-discussed arbitrary values of drilled pier end-bearing and skin friction capacity, geotechnical engineers in the Rocky Mountain area established a rational design method for piers.

Experience with shale bedrock in the Denver area generally indicates that the hardness of bedrock increases with depth (Figure 10.7). This has been verified by penetration resistance and by pressuremeter tests. Therefore, for piers penetrating deep into bedrock, a higher load-carrying capacity can be assigned. Experience indicates that both the end-bearing capacity and the skin friction value increase with depth at the rate of 3% per foot. Using this assumption, the load-carrying capacity of a straight-shaft pier can be expressed as follows:

$$Q = A(E + 0.3\ EL) + 0.1\ (E + 0.3\ EL)\ CL$$

In which: Q = total pier carrying capacity (kips)
A = end area of pier (ft^2)
E = end-bearing capacity of pier (psf)
L = depth of penetration into bedrock (ft.)
C = perimeter of pier (ft.)

Figure 10.7 was prepared using the equation given, with the assumption that the end-bearing value at the top of bedrock is 50,000 psf. It should be pointed out that the above equation has its limitation, as noted:

1. The bearing capacity of the top of bedrock should be carefully evaluated. Often, the top of the shale is highly weathered and only a low value can be assigned. Sometimes, deep excavation will expose the top of the shale and the material will be subject to severe disintegration. Therefore, it is usual to assign a bearing-capacity value for that portion of pier 2 to 4 ft below the surface of bedrock and then increase the bearing value for additional penetration.
2. The increase of bearing capacity with depth of penetration has its limitations. Extensive experience in the Rocky Mountain area indicates that the absolute maximum of end-bearing value for piers in shale is 100,000 psf.

Using the above approach in designing drilled piers, the load-carrying capacity of a single 60-in.-diameter pier can easily reach 3000 kips with reasonable penetration. Such a load is of sufficient magnitude to accommodate most column loads of a high-rise structure.

10.3.1 Friction Piers

Where the bedrock is deep and the upper soils cannot be relied upon to support the foundation system, the use of friction piers can be advantageous. The following should be observed:

1. During subsoil exploratory drilling, special attention should be directed to detecting any soft layers within the entire length of the friction pier. If such a stratum is present, the use of friction piers should be reevaluated.
2. In a high water table area, the use of friction piers is seldom justified. Water reduces the skin friction value.
3. Skin friction value should not be assigned to at least the upper 3 ft of the pier. Surface water tends to soften the soil surrounding the upper portion of the pier.
4. The end-bearing value of a friction pier can be disregarded and the skin friction value calculated in accordance with the shear strength of clay.
5. The use of a friction pier in granular soil or caving soil is seldom justified. If the soils are granular spread-footing, the foundation system should be considered.

10.3.2 Belled Piers

Piers drilled into materials other than bedrock are often enlarged at the bottom of the hole for the purpose of increasing the bearing area, thus increasing the total load-carrying capacity. Such piers are commonly referred to as belled piers, or underreamed piers. The ideal bell is in the shape of a frustum with vertical sides at the

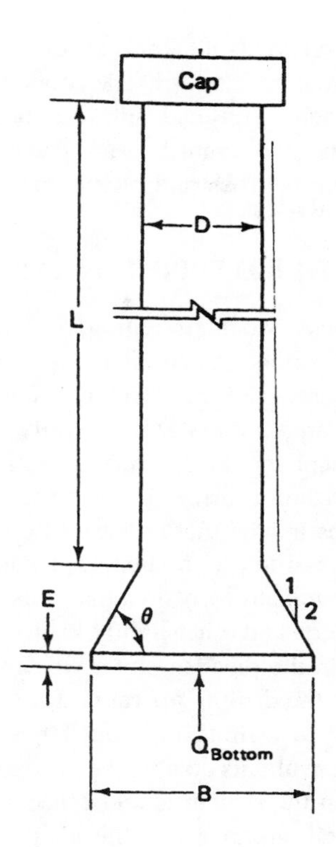

FIGURE 10.8 Typical Belled Pier.

bottom. The vertical sides are possibly 6 to 12 in. high. The sloping sides of the bell should be at an angle of at least 60° with the horizontal as shown in Figure 10.8. Most drillers are capable of providing bells with diameters equal to three times the diameter of the shaft.

Only the straight part of the shaft should be considered in computing skin friction capacity. The width is taken as the diameter of the bell in all determinations of the ultimate and safe pressures, and for the computation of the area and total bearing capacity of the pier.

The most prominent disadvantages of belled piers are the cost and the difficulty of inspection. Bells can be formed mechanically with a special belling device. The following should be considered in using the bell system:

1. When it is necessary to send a man down the hole with a jack hammer to complete the bell, the cost can be prohibitive.
2. It is not possible to form a bell in granular soils without the risk of caving.
3. In order to allow for inspection, the diameter of the shaft should be at least 24 in.
4. Assuming the use of a 24-in. shaft, the maximum bell diameter should be 6 ft for granular soil with a bearing capacity on the order of 6 ksf. The

total load-carrying capacity of the belled pier is therefore only on the order of 225 kips. The economics of using bells is therefore questionable.

5. There is little advantage in belling the pier on shale bedrock or hard clay. The additional bearing capacity can be easily offset by drilling deeper into bedrock and picking up the required bearing value in skin friction.

10.4 DRILLED PIER IN EXPANSIVE SOILS

In the Rocky Mountain region, drilled pier foundations have been used widely since 1950 to combat structural movement problems in expansive soils. It is now an accepted practice for geotechnical engineers to use drilled pier foundations in expansive soil areas. Piers are generally drilled into bedrock in a zone unaffected by moisture change. With some exceptions, such a system has successfully prevented damage from occurring in the structures as the result of soil expansion.

It is estimated that at least 70% of all piers drilled in the Rocky Mountain area can be classified as related to expansive soil problems. The use of such piers is generally related to residential construction, commercial buildings, schools, shopping centers, and other lightly loaded structures. The pier diameter varies from 10 to 24 in. with the pier length varying from 10 to 30 ft.

Since the drilling operation for such small-diameter piers is relatively simple as compared to drilling of deep, large-diameter caissons, both the drilling company and the consultants do not give the operation the attention it deserves. When problems develop in the structure, and especially when damage claims have been brought to court, questions arise as to the adequacy of the pier installation. Only then are records studied, inspectors questioned, and designs reviewed. The issues commonly brought forth are as follows:

Have the piers been drilled to the specific depth?
Has the reinforcement been properly installed?
Have mushrooms been present on the top of the piers?
Have the piers been placed directly beneath the grade beams?
Were there gaps or voids in the concrete?
Were the bottoms of the pier hole properly cleaned?
Who inspected and approved the pier installation?

These and other similar questions are repeatedly brought up by attorneys in court or at deposition. Lengthy legal disputes involving very costly procedures of excavating the entire length of the pier for examination and verification are commonly conducted. It is difficult to determine the cause of movement and even more difficult to ascertain responsibility.

Much of the litigation could be avoided if only the client understands the risk involved in the use of drilled piers, if only the contractor and the engineer better document the pier drilling operation, and if only the inspector and the operator pay more attention to the installation.

10.4.1 Swelling Pressure

Essentially, swelling pressure is required to maintain the initial soil volume when subjected to moisture increase. Swelling pressure can be determined in the laboratory under volume-controlled conditions. It can also be determined by allowing the soil to expand to its maximum volume, and then by measuring the pressure required to return it to its initial volume. Swelling pressure is a built-in soil property, varying only with the density of the soil. In the Rocky Mountain area, the swelling pressure of expansive clays varies from a few pounds per square foot to as high as 100,000 psf. Most of the high-swelling clays or claystone belong to the high-density Pierre shale.

The portion of swelling pressure that can be responsible for pier uplift has not been clearly defined. In 1974, it was estimated that only 15% of the swelling pressure can be taken as uplift along the pier shaft. This value agrees somewhat with the finding of Michael O'Neill's theoretical study in 1980. It is assumed that, for instance, if the pier is drilled into a shale bedrock with a swelling pressure of 20,000 psf, the uplifting pressure along the pier shaft is on the order of 3000 psf. This estimate must be verified before a rational pier design can be established.

10.4.2 Depth of Wetting

The theory that leads to the stability of the drilled pier foundation in expansive soils assumes that the sources of soil wetting are derived from the surface and gradually penetrate into the subsoil. The source of surface water that seeps into the subsoil can be from lawn irrigation, roof drains, melting snow, precipitation, broken water and sewer lines, and other sources. The most frequent and easy access for water to enter the foundation soil is through the loose backfill around the building.

Depth of wetting depends on the topographic features, the climatic conditions, and the type of subsoil. In the case where surface drainage is poor and watering is excessive, the depth of wetting can exceed 10 ft. Sometimes the entire length along the pier can be wetted.

There is no reliable information on the state of wetting that claystone shale can receive from surface wetting. Much depends on the pattern of the seams and fissures and the water-carrying sandstone lenses. Assuming that surface irrigation has been controlled and the ground surface slopes away from the structure in all directions, the surface water should not penetrate into the clay soil for more than 5 ft.

A perched water condition can develop over several years after the completion of a subdivision or a commercial area for several years. Perched water travels in the seams and fissures of shale. Since the interface between the pier and the soil have more affinity with water than the remainder of the soil mass, it is almost certain that the pier will puncture the seams and allow water to enter the space surrounding the pier. As soon as this takes place, the skin friction is lost and swelling along the dry portion of the pier takes over, resulting in pier uplift.

A drilled pier foundation system will no longer be safe should a perched water condition develop in the area. The commonly accepted concept that the longer the penetration of pier into bedrock, the safer the system, may not be true in this case.

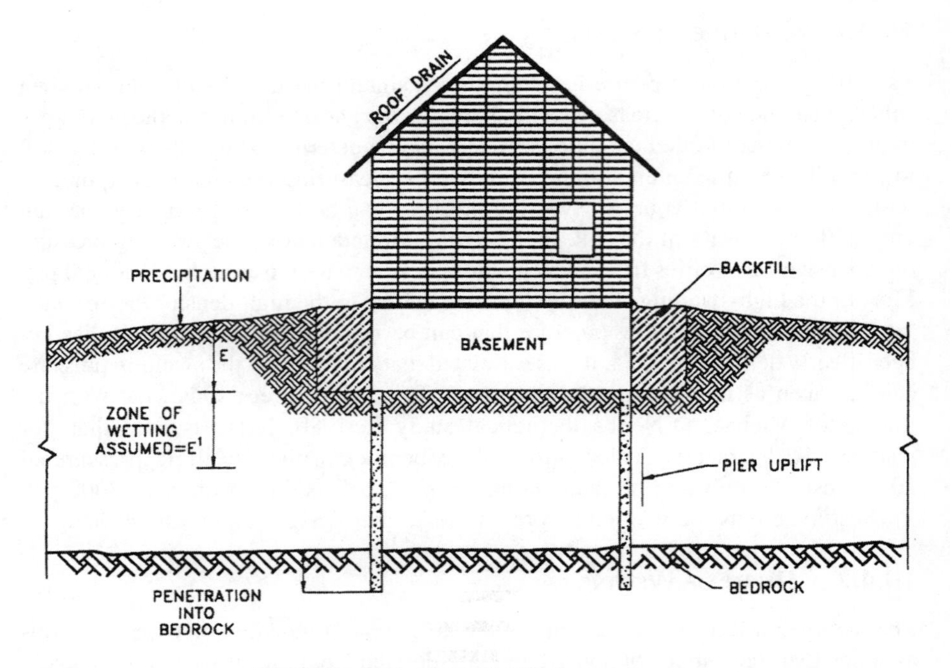

FIGURE 10.9 Typical drilled pier system.

In fact, it may even have an adverse effect. The use of a drilled pier foundation system in an area where perched water is present requires careful evaluation.

10.4.3 Design Criteria

The design criteria for a drilled pier foundation system in an expansive soil area can be illustrated by using Figure 10.9 with a typical calculation:

1. It is assumed that the overburden clays are expansive with a swelling pressure of 20,000 psf and a skin friction of 1500 psf. It is assumed that 15% of the swelling pressure is transmitted to the pier, causing pier uplift. Claystone bedrock with a penetration resistance of $N = 50$ is located at a depth 10 ft below the basement floor level. Bedrock has a natural moisture content below 10% and contains seams and fissures with isolated sandstone lenses.
2. Drainage around the building is in fair condition. Roof and surface water are able to drain away from the structure in all directions. However, backfill around the building is loose. Water is able to accumulate in the backfill and seep into the natural soils. The depth of wetting is assumed to be 5 ft.
3. A drilled foundation system is used. The pier diameter is 12 in. and penetrates about 5 ft into the claystone bedrock. The total length of the pier is 15 ft.
4. Piers are spaced 15 ft apart, with an estimated dead load on each pier of 20 kips.

$$\begin{aligned}\text{Total uplifting force} &= (\text{pier perimeter}) \times (\text{depth of wetting}) \times (\% \text{ of uplift}) \\ &= 3.14 \times 1 \times 5 \times 20 \times 0.15 \\ &= 47.1 \text{ kips}\end{aligned}$$

$$\begin{aligned}\text{Total withholding} &= (\text{dead load pressure}) + (\text{pier perimeter} \times \text{skin friction} \\ &\qquad \times \text{pier penetration}) \\ &= 20 + 3.14 \times 1 \times 1.5 \times 5 \\ &= 43.6 \text{ kips}\end{aligned}$$

The above simple calculation indicates that the total withholding force is approximately equal to the total uplifting force, and the pier system is safe. No factor of safety has been assigned in the calculation. It is taken for granted that the various assumptions have a built-in factor of safety.

This typical method of calculation for drilled pier foundations has been accepted with reasonable success by geotechnical and structural engineers in the area for many years. In residential construction, there is no difficulty in meeting the minimum dead-load pressure requirement on the order of 20,000 psf. However, in the basement portion of the house, under the steel posts supporting the central beam and also the patio posts, there will be difficulties in meeting the dead-load pressure requirements. The solution is in accordance with the above calculation to increase the withholding force by penetrating deeper into the bedrock.

Geotechnical engineers should realize that the answer of how to found lightly loaded structures on expansive soils has not been reached. A great deal of research and field tests are required before a reasonable solution can be applied.

10.4.4 Belled Pier

Probably the greatest advantage of the belled pier is that the resistance against uplift is not affected by the loss of friction in the zone unaffected by wetting. Since the withholding force depends on the friction of the pier, if the friction is lost, then pier uplift is unavoidable. The loss of friction can easily be caused by the rise of ground water or the development of a perched water condition. The total weight of the soil above the bell will not be affected by moisture change. Hence, there is always an added factor of safety of the belled pier system against pier uplift.

Since a belled pier system relies entirely upon the anchorage of the lower portion of the pier in the zone unaffected by wetting against uplift, this type of pier can be used for columns supporting very light loads. The magnitude of dead-load pressure exerted on the pier is not a factor.

10.5 PIER CONSTRUCTION

Many leading drilling companies with the capability of drilling piers up to 6 ft in diameter are stationed in Denver. The technique of pier drilling has improved steadily. Difficult projects involving high water tables, caving soils, large boulders, and other problems have been successfully overcome by experienced drillers. It can be safely stated that no project has been abandoned due to the failure of the driller to complete the pier drilling operation.

10.5.1 Pier Hole Cleaning

To ensure that the bottom of the pier hole is clean and free of loose earth, the pier hole must be properly cleaned. This can usually be accomplished by adding a small amount of water into the pier hole and spinning the auger lightly so that the loose earth will adhere to the auger and be removed. If loose rocks and soft mud are present in the bottom of the hole, it may be necessary to send a helper down the hole to clean the hole by hand. Such an operation is possible only for large-diameter piers. Failure to clean the bottom of the pier hole can sometimes result in excessive settlement. Fortunately, most of the small-diameter piers are overdesigned, and the skin friction alone is sufficient to support the column load. The condition of the pier bottom is therefore not as critical.

10.5.2 Dewatering

Groundwater can seep into the pier hole through the upper overburden soils or through the lower bedrock. A pier-drilling operation usually seals the seams in the soil and stops seepage temporarily. Consequently, if concrete is available at the site and poured immediately after the completion of the drilling, dewatering can be avoided. On the other hand, if the pier hole is allowed to stand for a long period of time, water will seep into the hole and must be pumped out before pouring concrete. Pier holes left open for a long time can also result in costly hole remediation. An experienced driller, under such conditions, would rather fill up the hole and re-drill the hole when concrete is available for immediate pouring.

If water enters into the pier hole rapidly through the upper granular soils, it will be necessary to case the hole above bedrock to control seepage. In most instances, the use of casing will seal all seepage through the overburden soils. However, casing above bedrock cannot stop the infiltration of water through the seams and fissures of the bedrock. Such water must be pumped out. If water is mixed with auger cuttings in a form of slurry, such a mixture can be bailed out by the use of a bailing bucket.

10.5.3 Concrete in Water

Specifications generally call for the pouring of concrete in less than 6 in. of water in the pier hole. In fact, concrete can be poured successfully in less than 12 in. of water. Concrete displaces water and forces water to the top of the pier hole, where it drains away. If it is necessary to pour concrete in deep water, a tremie should be used. The bottom of the tremie should be kept below the surface of the concrete. Concrete is introduced into the hole by the use of an elephant trunk or by pressure pumping to avoid the effect of the segregation of concrete. If concrete is not allowed to hit the wall of the drill hole, high free fall of as much as 100 ft will not cause segregation.

10.5.4 Casing Removal

Steel casings are costly, and whenever possible the driller removes the casing after the completion of the pouring. Hasty removal of the casing can introduce air pockets in the pier shaft that eventually will be filled with surrounding soils. This is especially serious where the pier is heavily reinforced. The John Hancock Building in Chicago suffered considerable construction delays due to poor piers that resulted from hasty casing removal. In an army project in Colorado, a 24-in. diameter pier reinforced with six 3/4-in. bars in a cage settled more than 3 ft even before the load was applied to the column.

Under some difficult circumstances, it is prudent to leave the casing in rather than expend the effort to remove it. However, in such cases, the engineer should be aware of the loss of skin friction with a smooth casing surface.

Direct observation of the elevation of the top of concrete is difficult. If the surface of the concrete rises even momentarily as the casing is being withdrawn, it is virtually certain that the pier hole will be invaded by the surrounding soils or foreign material. The appearance of a sinkhole or depression of the ground surface near the pier hole also offers a good indication of faulty installation. It is also necessary to compare the volume of concrete poured with the volume of the pier hole. Defects in the pier shaft due to casing removal can only be prevented by an experienced driller and a field engineer with powers of keen observation.

10.5.5 Specification

Construction specifications sometimes are prepared by engineers with no field experience who copy from some previous project. Errors related to concrete slump, aggregate size, pier diameter, etc. are found in some hastily prepared specifications. An experienced contractor would point out these problems before the commencement of the project. They would rather stop the work than adhere blindly to the specification.

10.5.6 Angled Drilling

In some unusual cases, it is necessary to drill pier holes at an angle (Figure 10.10). Near-horizontal drilling has been attempted. Such piers can be used as a tie-back for retaining walls or sheeting for deep excavation. The mechanics of such piers is seldom reported. The design as well as the method of construction should be carefully studied by both structural and geotechnical engineers.

10.6 PIER INSPECTION

For a geotechnical engineer, more important than any of the above theoretical approaches to pier analysis is the pier inspection. Excessive settlements of the piers resulting in building distress generally are not caused by faulty analysis or errors in design, but by defective pier construction. The more common problems resulting in defective piers are:

10.6.1 Regulations

Prior to 1960, there were no regulations on the safety requirements for pier inspection. Engineers rode on the driller's kelley bar descending into an uncased drill hole.

FIGURE 10.10 Pier drilled at an angle.

FIGURE 10.11 Skyline of downtown Denver, buildings founded on drilled pier.

It was quite an experience for those in a deep bore hole, looking up at the sky which appeared to be the size of a dime. Although the risk involved in entering an uncased hole is large, accidents are seldom reported.

Today, OSHA has strict regulations on pier inspection. The commonly accepted rules are:

Never enter an uncased drill hole.
Always wear a harness with a safety device. This is to guard against attack by noxious gas.
An inspector should not enter holes with too much water. Water should be pumped out prior to entering.
Inspection procedures should be completed as quickly as possible. The entire operation should be completed in about 5 min to avoid delaying the pouring of concrete and the deterioration of the condition of a clean hole.

10.6.2 Pier Bottom

The cleaning of loose soil at the bottom of the pier hole after the pier drilling has reached the design depth is important to prevent undue pier settlement. For small-diameter or uncased piers, the holes can be inspected by shining a mirror or a strong light into the hole. If the hole is not too deep, a fair evaluation can be achieved. For deep holes, above-ground inspection is not adequate; it is necessary to enter the pier hole and visually evaluate the condition.

The presence of as little as 1 in. of loose soil at the bottom of the shaft can cause unacceptable settlement. Geotechnical consultants in Pierre, South Dakota, specified that all deep pier holes must be inspected by the use of a hand penetrometer.

Inspection of pier holes becomes difficult when water is present. The engineer should try to enter the pier hole immediately after the completion of the drilling and before any seepage has built up. In some cases, it will be necessary to pump the water out before entering.

If the settlement due to loose soil at the bottom of the pier is not excessive, it can be corrected by shimming the pier top. However, care should be taken to ensure that all settlement has taken place. Oftentimes, total settlement will not take place until the structure is completed and occupied.

10.6.3 Pier Shaft

Concrete can adhere to the wall of the shaft, creating a large void along the wall and preventing the concrete from dropping to the bottom of the shaft. For large-diameter piers, such occurrences are usually caused by large aggregates logged between the shaft and the steel cage. For small-diameter piers, the adhesion between concrete and shaft also can prevent the concrete from reaching to the bottom.

Such occurrences generally are caused by using either too large an aggregate or too stiff a concrete. Architects sometimes use the same specifications of concrete for piers as for slabs and beams. As a result, the use of low-slump concrete and oversize aggregates causes the problem. It is always desirable for the geotechnical engineer to review the foundation specification before entering the bid.

Checking the volume of the drill hole with the amount of concrete actually used can sometimes reveal the error. However, most of the time the defective piers can temporarily be held up by skin friction and are not detected until the building load is applied. Settlement of the pier by as much as several feet has been reported. This can be very serious and very difficult to correct.

In one Corps of Engineers project, it was necessary to build platforms on top of each pier and drill holes through the piers to detect the condition of the pier bottom.

The diameter of the piers in expansive soils should be as small as possible, in order to concentrate the dead load to prevent pier uplift. Experience indicates that piers smaller than 12 in. are difficult to clean. It is recommended that all piers drilled in expansive soils should have a diameter no less than 10 in.

REFERENCES

F.H. Chen, *Foundations on Expansive Soils,* Elsevier Science, New York, 1988.

P.M. Goeke and P.A. Hustad. Instrumented Drilled Shafts in Clay-Shale, presented at the October ASCE Convention and Exposition, Atlanta, GA, 1979.

W.R. and W.S. Greer, *Drilled Pier Foundations,* McGraw-Hill, 1972.

R.G. Horvath and T.C. Kenney, Shaft Resistance of Rock-Socked Drilled Piers, presented at the ASCE Convention and Exposition, Atlanta, GA, 1979.

D. Jubervilles and R. Hepworth, Drilled Pier Foundation in Shale, Denver, Colorado Area, Proceedings of the Session on Drilled Piers and Caissons, ASCE/St. Louis, MO, 1981.

M.W. O'Neill and N. Poormoayed, Methodology for Foundations on Expansive Clays, Journal of the Geotechnical Engineering Division, ASCE, Vol. 106, No. GT 12, 1980.

H.G. Poulos and E.H. Dais, Settlement Analysis of Single Piles, *Pile Foundation Analysis and Design,* John Wiley & Sons, New York, 1980.

W.C. Teng, *Foundation Design,* Prentice-Hall, Englewood Cliffs, NJ, 1962.

11 Laterally Loaded Piers

CONTENTS

The design of laterally loaded piers drilled in cohesive soils and cohesionless soils has been investigated by many authors in the late 20th century. In 1955, Terzaghi used subgrade reaction as the criteria for the design of lateral load on piles. In 1957, Czerniak made exhaustive structural analysis on long and short piles based on Terzaghi's suggested values of subgrade reaction. Computer programs were set using Peter Kocsis', Reese's, or Matlock's analysis. Such programs have been used by many consulting structural engineers. The shortcoming of such analysis is that while the structural analysis is elaborate, the main source of input on soil behavior is foggy and very sketchy.

Probably the most complete review on the design of laterally loaded piles was given by B.B. Broms in 1965. Broms covered this design in his investigation on long and short piles both in free ends and in restrained condition, in cohesive and in cohesionless soils. His analysis is based on both the conceptions of lateral soil resistance and on lateral deflection. By following Broms' reasoning and by inserting the actual subsoil conditions and drilled pier system in the Rocky Mountain area, a rational laterally loaded pier design procedure can be established.

By following the charts and figures in this chapter, the consultant will be able to assign values of lateral pressure without entering into lengthy calculations.

11.1 DESIGN CRITERIA

The behavior of a laterally loaded pier depends on many parameters. Some of the more important ones are discussed as:

11.1.1 Degree of Fixity

The behavior of a laterally loaded pier depends on the degree of fixity imposed at the top of the pier by the supporting structure. A pier system supporting high-rise structures can generally be considered to be a fix-head or in a restrained condition. Such piers are subject to wind load, earth pressure, or earthquake load. The deflection criterion generally controls the design.

Free-headed piers are those of transmission towers, sign posts, light poles, etc. Such structures can tolerate large deflection, and the maximum soil resistance controls their design. In this review, more attention is paid to the free-headed condition, since if the foundation system is safe for the free-headed condition then, under the restrained condition, the factor of safety will be ample.

11.1.2 Stiffness Factor

The flexural stiffness of the pier relative to the stiffness of the material surrounding the upper portion of the shaft also controls the pier behavior. The loads against the deflection characteristics of a "rigid" pier are, therefore, quite different from those of an "elastic" pier.

The demarcation between elastic and rigid pier behavior can be determined in terms of a relative stiffness factor that expresses a relation between soil stiffness and pier flexural stiffness. For cohesive soil, the stiffness factor is:

$$\beta = \left[K_I D/4\ E\ I\right]^{1/4}$$

in which k_l = Coefficient of horizontal subgrade reaction (tons/ft^3 or lbs/in.^{3})
D = Pier diameter (in.)
E = Modulus of elasticity of concrete (lbs/in.^{2})
I = Moment of inertia of pier section (in.^{4})

When β L is larger than 2.25, a long and elastic pier condition is assumed, and when βL is less than 2.25, a short and rigid pier behavior can be assumed. L is the length of the pier embedment.

For a pier drilled into cohesionless soils, where the soil modulus increases linearly with depth, the stiffness factor is

$$\mu = \left[n_h/E\ I\right]^{1/5}$$

in which n_h is the constant of the horizontal subgrade reaction for piers embedded in sands.

Where μL is larger than 4.0, the pier is treated as an infinitely long, elastic member, and when μL is smaller than 2.0. a short rigid pier is assumed.

In this chapter, the diameter of the pier ranges from 18 to 42 in. Problems of lateral load are generally associated with pier diameters in this order. Depth of embedment is limited to 10, 20, and 30 ft. Contrary to structural piles, the L/D ratio of a pier is generally much smaller than a pile. Lateral deflection will not be affected by pier length when an elastic pier behavior has taken place.

11.1.3 Surrounding Soils

Two major categories of soils surrounding the piers are cohesive soils and cohesionless soils. Cohesive soils include clays, sandy clays, and silty clays of various consistencies. They also include weathered claystone and claystone bedrock. Medium-hard clays with unconfined compressive strength on the order of 8000 psf generally belong to the weathered bedrock category, while all bedrock has unconfined compressive strength of at least 15,000 psf.

For cohesionless soil, consideration should be given to the grain size. Soil with a high percentage of gravel is generally high in relative density, in unit weight, and in friction angle, as compared to soil with a low percentage of gravel.

The upper soils surrounding the pier govern the behavior of the pier under lateral pressure. For example, if a pier is drilled through 10-ft dense sand and gravel into bedrock, the cohesionless soil controls the magnitude of allowable lateral pressure, not bedrock. At the same time, if a pier is drilled through a thin layer of sand or soft clay into bedrock, then the embedment of the pier in bedrock controls the allowable lateral pressure. Detailed analysis of stratified soil at various depths, either by graphic method or by computer, is not warranted.

11.1.4 Movement Mechanics

The pier under a lateral load pivots about a point somewhere along its length (about 3 pier diameters). As resistance to the applied loading is developed, the soil located in front of the loaded pier close to the ground surface moves upward in the direction of least resistance, while the soil located at some depth below the ground surface moves in a lateral direction from the front to the back side of the pier. At the same time, the soil separates from the pier on its back side to depth below the ground surface as shown in Figure 11.1.

The design of a laterally loaded pier is, in general, governed by the requirements that complete collapse of the pier should not occur even under the most adverse conditions and that the deflections or deformations at working load should not be so excessive as to impair the proper function of the foundation.

Thus, for the type of structure in which small lateral deflections can be tolerated, the design is governed by the lateral deflection at working loads. The deflection of a laterally loaded pier can at working loads be calculated based on the concept of coefficient of subgrade reaction. The ratio of the soil reaction and the corresponding lateral deflection is either constant or increases linearly with depth.

For structures in which a relatively large deflection can be tolerated, the design is governed by the ultimate lateral resistance of the pier. Ultimate lateral resistance of a relatively small embedment is governed by the passive lateral resistance of the soil surrounding the piers.

The soil information can be obtained by unconfined compressive strength tests on cohesive soils and the direct shear test on cohesionless soil.

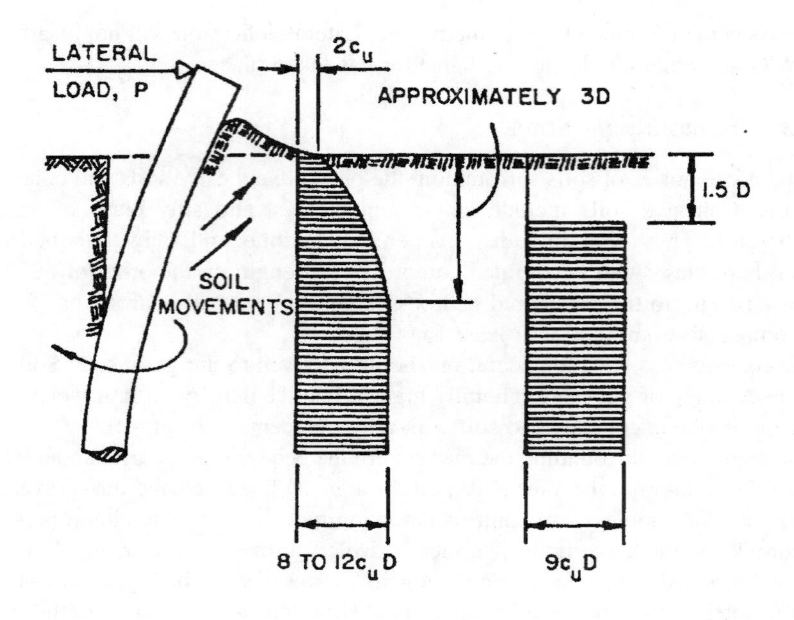

FIGURE 11.1 Distribution of lateral earth pressures in cohesive soils (after Brom).

11.2 LIMITING CONDITIONS

In applying various data obtained from both laboratory and the field, certain modifications and refinements are required:

Shape Factor — Since the curved surface of a pier can penetrate the earth more easily than the flat surface, the effectiveness of a round pier must be decreased. A shape factor of 0.8 is recommended. Thus, in considering the resistance of a round pier, the effective width may be taken as 0.8 of the pier diameter.

Under Strength Factor — The ultimate lateral resistance may be calculated on the basis of reduced cohesive strength, equal to the under strength factor times the measured or estimated cohesive strength. The design cohesive strength may be taken as 75% of the minimum measured average strength within the significant depth.

Pier Surface — The ultimate lateral resistance of a pier embedded in clay varies with the condition of the pier surface. Apparently, lateral resistance is greater for a rough pier surface than a smooth one, such as steel.

Repetitive Loading — Repetitive loading can decrease the ultimate lateral resistance of cohesive soil to about half its initial value. Repetitive loading and vibration may cause substantial increase of the deflection in cohesionless soils, especially if the relative density of the surrounding soils is low.

Load Factor — The lateral forces acting on a pier caused by earthquakes, waves, or wind forces are frequently difficult to calculate or to estimate. High load factors should be used when the applied load can be estimated accurately. Frequently, a load factor of 1.50 is used with respect to a live load.

Allowable Deflection — Allowable deflection varies considerably with different types of structures. For tower structures, such as transmission towers, antennas, sign posts, and others, a large deflection on the order of several inches can be tolerated. For high-rise structures, the structural engineer generally calls for maximum lateral deflection at the top of the piers not to exceed 0.25 in.

In the deflection analysis (Figure 11.2), two criteria have been used:

1. Use free-headed piers with a maximum deflection of 0.5 in.
2. Use fixed-headed piers with maximum deflection of 0.25 in.

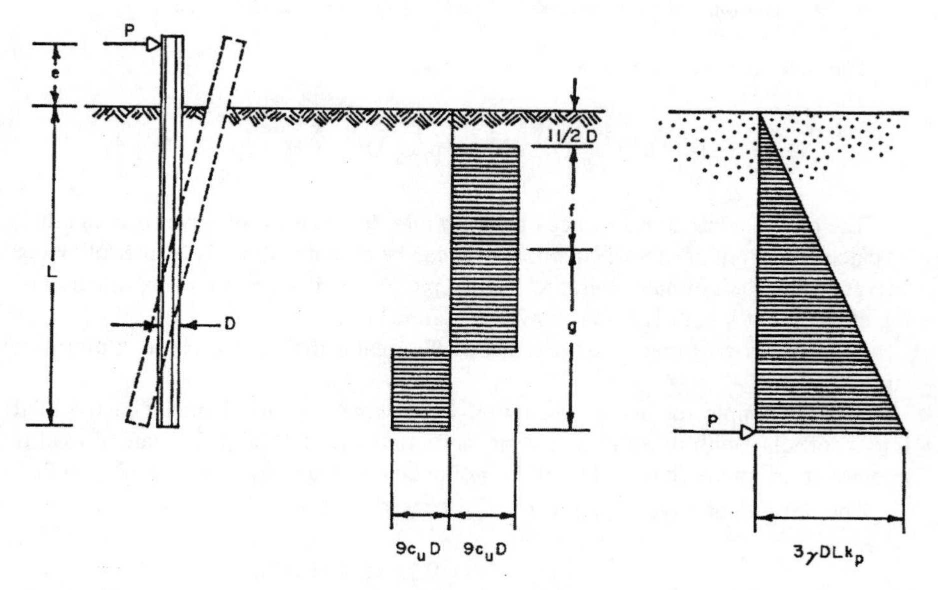

FIGURE 11.2 Assumed distribution of lateral earth pressure at failure of a free-headed pier drilled in cohesion or cohesionless soil (after Broms).

11.3 ULTIMATE LATERAL RESISTANCE OF COHESIVE SOILS

When the ultimate soil resistance is reached, the body of the soil is on the verge of failure. The calculated value of this resistance or passive pressure may be used for design purposes. The lateral earth pressure acting at failure on a laterally loaded pier in a saturated cohesive soil is approximately 2 c_u at the ground surface, in which c_u is the cohesive strength as measured by unconfined compressive or vane tests. The lateral soil reactions increase with depth and reach a maximum of 8 to 12 c_u at approximately 3 pier diameters below the ground surface, as shown in Figure 11.2. The lateral soil reactions may be assumed equal to zero down to a depth of 1.5 pier diameters and equal to 9 c_u D below this depth.

The maximum moment occurs at a level where the total shear force in the pier is equal to zero at a depth (f +1.5D) below the ground surface, as shown in Figure 11.2. The distance f and the maximum bending moment M can be calculated from the following equations:

$$f = p/9\ c_u\ D$$

$$M = p(e + 1.5\ D + 0.5\ f)$$

$$M = 2.25\ D\ 9^2 \quad L = (1.5\ D + f + g)$$

Where e = eccentricity of the applied load
g = the part of pier located below the point of maximum bending

The resulting equation in terms of L/D is:

$$L/D = 1.5 + 0.111\ (p/c_u\ D^2) + \left[0.66(p/c_u\ D^2) + 0.024\ (p + c_u\ D^2)^2\right]^{1/2}$$

The ultimate lateral resistance of a short pier drilled into cohesive soils can then be calculated from the above equation, or it can be obtained directly by the following curves where the ultimate lateral resistance p/c_uD^2 has been plotted as a function of the embedment length L/D, as shown in Figure 11.3.

It is seen from Figure 11.3 that with L/D greater than 5, the curve is almost a straight line.

As an example for using this figure, we assume a 30-in. diameter pier drilled into a soft clay with unconfined compressive strength of 1000 psf. A lateral load is applied at ground surface, e/D = 0, D = diameter × shape factor = (2.5)(0.8) = 2.

The design cohesive strength $C_{design} = 0.75\ Cu$

$$= (0.75)(0.5)(1{,}000)$$

$$= 375\ \text{psf}$$

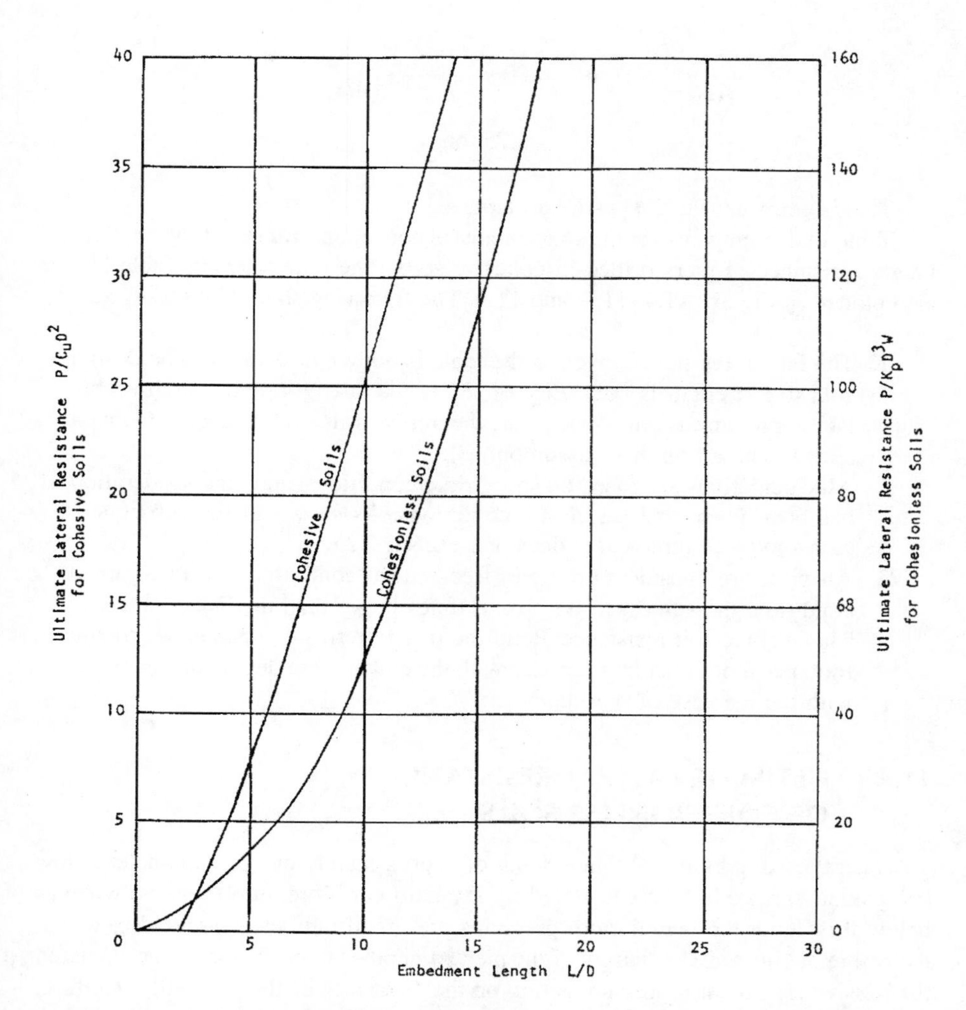

FIGURE 11.3 Ultimate lateral resistance to embedment length for free-headed piers drilled in cohesionless and cohesive soils.

$$\frac{p_{design}}{C_{design}\ D^2} = \frac{2p_{calc}}{375\ 2^2}$$

with total embedment of 20 ft, L/D = 8, from Figure 11.3,

$$\frac{p}{Cu\ D^2} = 18$$

$$p_{calc} = \frac{18 \times 375 \times 2^2}{2}$$

$$= 6.75 \text{ tons}$$

This is equivalent to 270 psf/ft of depth.

Table 11.1 is prepared for the convenience of consultants for selecting the ultimate lateral resistance of piers drilled in cohesive soils. The data listed in Table 11.1 is also plotted in Figures 11.4, 11.5, and 11.6. The following should be observed:

1. The lateral resistance given in the table is quite conservative. The design cohesive strength is only 75% of the tested strength, and the allowable lateral pressure is only 50% of the theoretical value (the calculation given above is based on these assumptions).
2. All calculations are based on short, rigid conditions, and no consideration has been given to long elastic conditions. As shown in the subsequent tables, only in three cases does β L exceed 2.25.
3. All piers are considered to be in free-headed condition. For a restrained condition, the value will be several times larger than the listed values.
4. The average soil resistance along the pier in terms of pounds per square foot, per foot of depth, increases with the embedment depth and decreases with the increase of pier diameter.

11.4 ULTIMATE LATERAL RESISTANCE OF COHESIONLESS SOILS

The lateral earth pressure within a depth of approximately one pier diameter below the ground surface can be calculated by standard earth pressure theories, whereas below this depth the lateral earth pressures are greatly affected by arching within the soil in the immediate vicinity of the pier. At depths larger than one pier diameter, the passive lateral earth pressure acting on the front face of the pier will, therefore, at failure, considerably exceed the Rankine passive pressure, while the lateral earth pressure acting on the back face of the pier will be considerably smaller than the active Rankine earth pressure.

Lateral earth pressure at failure can be safely estimated as three times the passive Rankine earth pressure. The assumed distribution of lateral earth pressure at failure is shown in Figure 11.2. At depth z below the ground surface, the assumed soil reaction p per unit length of the pier will be:

$$p = 3\, D\, \gamma\, z\, K_p$$

in which γ = unit weight of the soil

K_p = Rankine earth pressure

$$= \frac{1 + \sin \phi}{1 - \sin \phi}$$

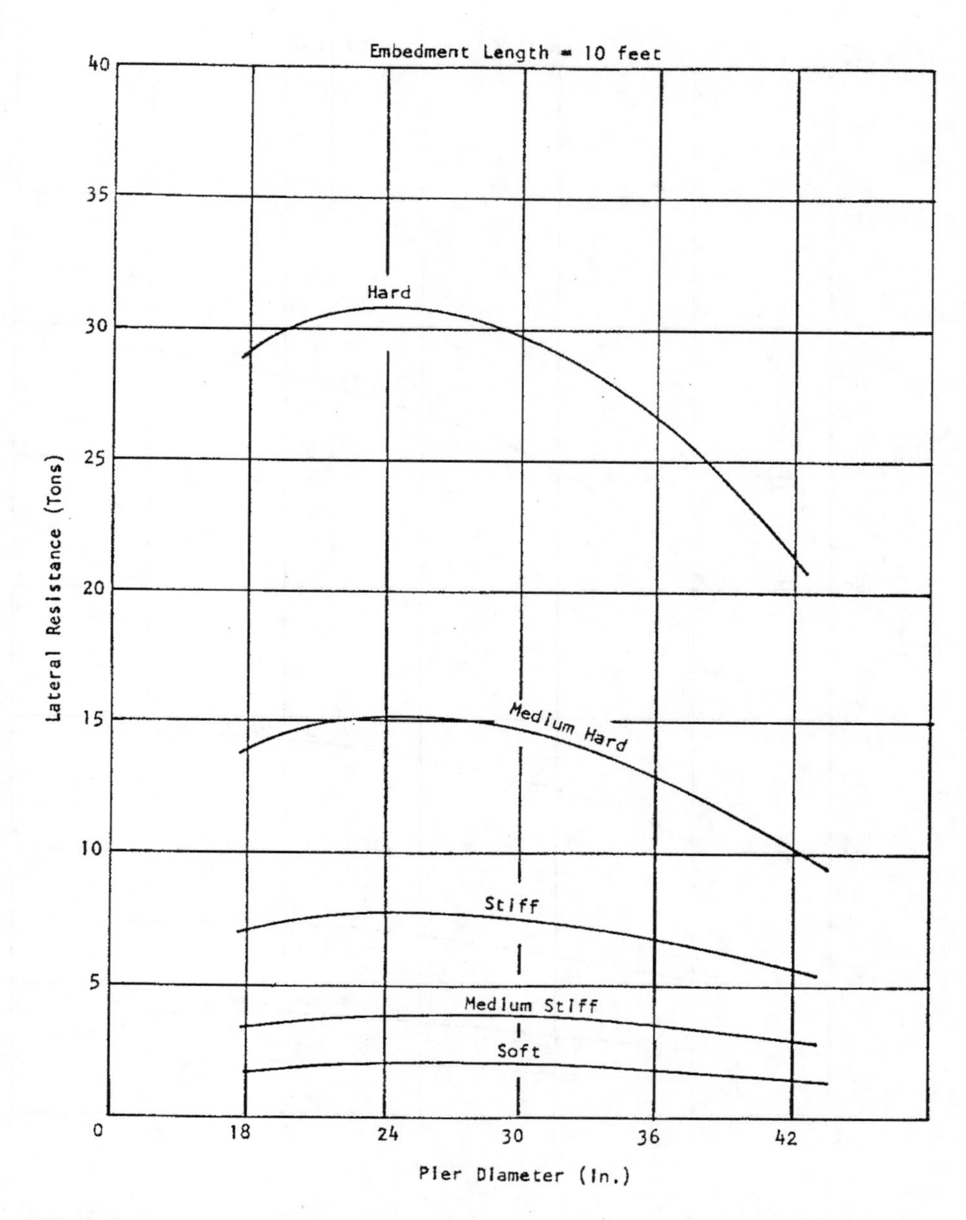

FIGURE 11.4 Lateral resistance of free-headed piers of various diameters drilled in cohesive soils.

The ultimate lateral resistance can then be evaluated from the equilibrium requirements. The driving moment M_d caused by applied load p with respect to the toe of the pier is:

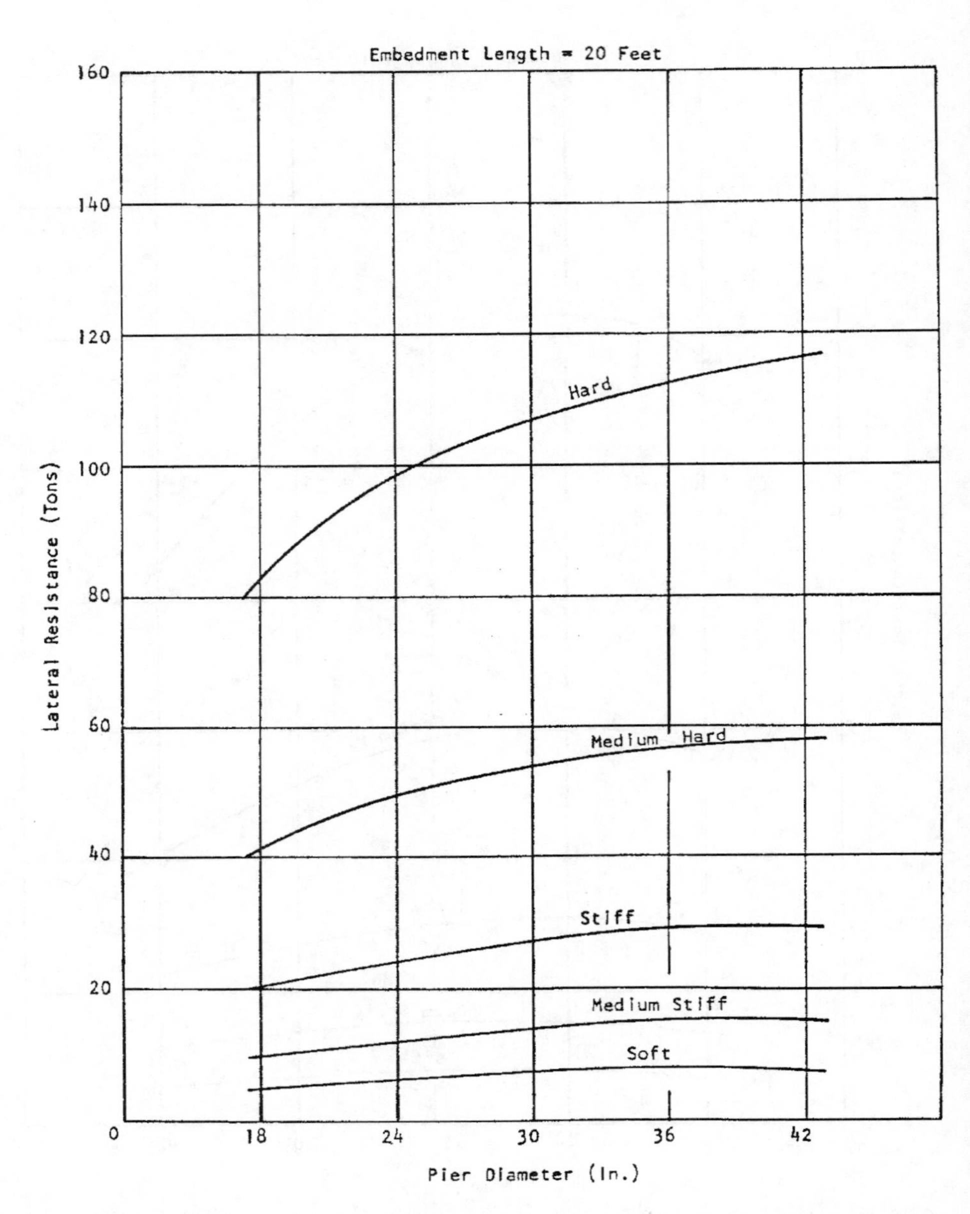

FIGURE 11.5 Lateral resistance of free-headed piers of various diameters drilled in cohesive soil.

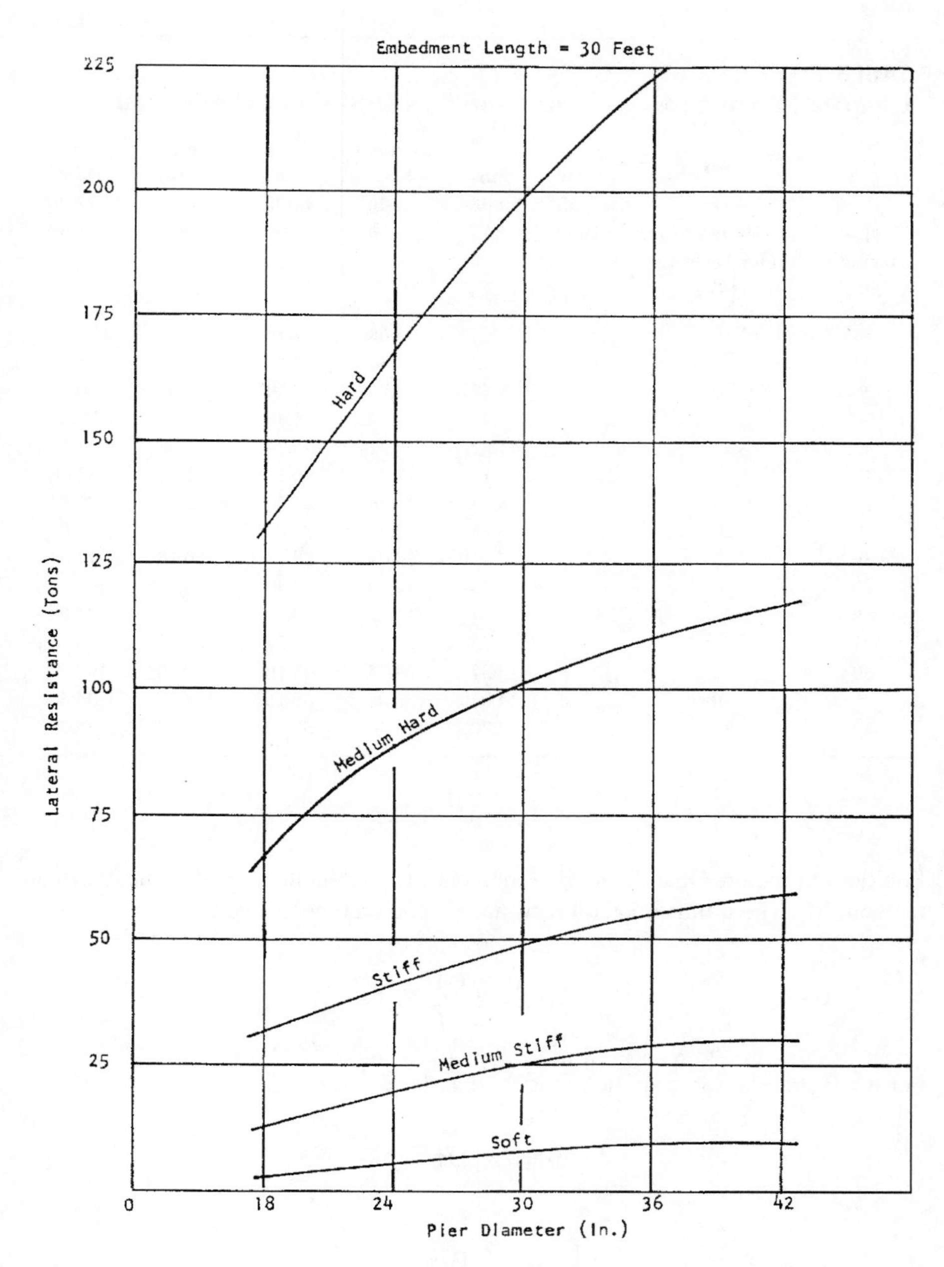

FIGURE 11.6 Lateral resistance of free-headed piers of various diameters drilled in cohesive soils.

TABLE 11.1
Ultimate Lateral Resistance (tons) for Piers Drilled in Cohesive Soils

	State	Soft	Medium Stiff	Stiff	Medium Hard	Hard
	Cu (psf)	1,000	2,000	4,000	8,000	15,000
Pier	Design Cohesion (psf)	375	750	1,500	1,000	5,600
Diameter (in.)	Pier Length (ft)					
18	10	1.84	3.68	7.36	14.72	29.44
24	10	1.92	3.84	7.68	15.36	30.72
30	10	1.87	3.75	7.50	15.00	30.00
36	10	1.62	3.24	6.48	12.96	25.92
42	10	1.32	2.92	5.24	10.56	21.12
18	20	5.13	10.26	20.52	41.04	82.08
24	20	6.24	12.48	24.96	49.92	99.84
30	20	6.75	13.50	27.00	54.00	108.00
36	20	7.02	14.04	28.08	56.16	112.32
42	20	7.35	14.70	29.40	58.80	117.60
18	30	8.37	16.75	33.50	67.00	134.00
24	30	10.56	21.12	42.24	84.48	168.96
30	30	12.37	24.75	49.50	99.00	198.00
36	30	14.04	28.08	56.15	112.30	224.60
42	30	14.70	29.40	58.80	117.60	235.20

$$M_d = 0.5\ D\ \gamma\ L^3\ k_p$$

Equilibrium requires that the driving moment M_d is at failure, equal to the resisting moment M_r. The ultimate lateral resistance p can then be evaluated as:

$$p = \frac{0.5 \gamma D L^3 k_p}{e + L}$$

For e = 0 (Lateral force applied at ground surface)

$$p = 0.5 \gamma D L^2 k_p$$

or

$$\frac{L}{D} = \left[\frac{p}{0.5\ k_p\ \gamma D^3} \right]^{1/2}$$

The dimensionless ultimate lateral resistance $p/k_p \gamma D^3$ determined from the above equation has been plotted in Figure 11.3 as a function of the dimensionless embedment length L/D.

As an example for using Figure 11.3, we assume a 30-in. diameter pier drilled into dense, cohesionless granular soils with an angle of internal friction of 40°. A lateral load is applied at ground surface, e/D = 0. D = (2.5)(0.8) = 2.0.

With total embedment of 20 ft, L/D = 8, from Figure 11.3

$$p/k_p \gamma D^3 = 32$$

$$p = (32)(4.60)(130)(2.0)^3$$

$$= 76.04 \text{ tons}$$

$$p_{calc} = \frac{76.04}{2} = 38.02 \text{ tons}$$

Table 11.2 is prepared for the convenience of consulting engineers in selecting the ultimate lateral resistance of piers drilled in cohesionless soils. The data listed in Table 11.2 is also plotted in Figures 11.7, 11.8 and 11.9.

TABLE 11.2
Ultimate Lateral Resistance (tons) for Piers Drilled in Cohesionless Soils

		State	Very Loose	Loose	Medium	Dense	Very Dense
		Relative Density (%)	20	30	50	70	80
		Penetration Resistance	5	10	20	40	50
		Friction Angle (Degree)	30	35	40	45	50
		Density (pcf)	120	125	130	140	145
Pier Diameter (in.)	Pier Length (ft)	L/D					
18	10	6.66	3.45	4.37	5.67	7.80	10.48
24	10	5.00	4.61	5.84	7.60	10.43	14.01
30	10	4.00	5.75	7.30	9.50	13.03	17.51
36	10	3.33	6.89	8.73	11.37	15.60	20.96
42	10	2.85	8.02	10.16	13.23	18.15	24.39
18	20	13.33	13.81	17.50	22.78	31.27	41.92
24	20	10.00	18.43	23.35	30.41	41.71	56.05
30	20	8.00	23.04	29.19	38.02	52.15	70.06
36	20	6.66	27.58	34.94	45.51	62.43	83.87
42	20	5.70	32.08	40.45	52.94	72.62	97.56
18	30	20.00	31.04	39.42	51.33	70.39	94.58
24	30	15.00	41.47	52.56	68.44	93.86	126.11
30	30	12.00	51.84	65.70	85.55	117.33	157.64
36	30	10.00	62.20	78.84	102.66	140.79	189.17
42	30	8.55	72.21	91.51	119.17	163.43	219.59

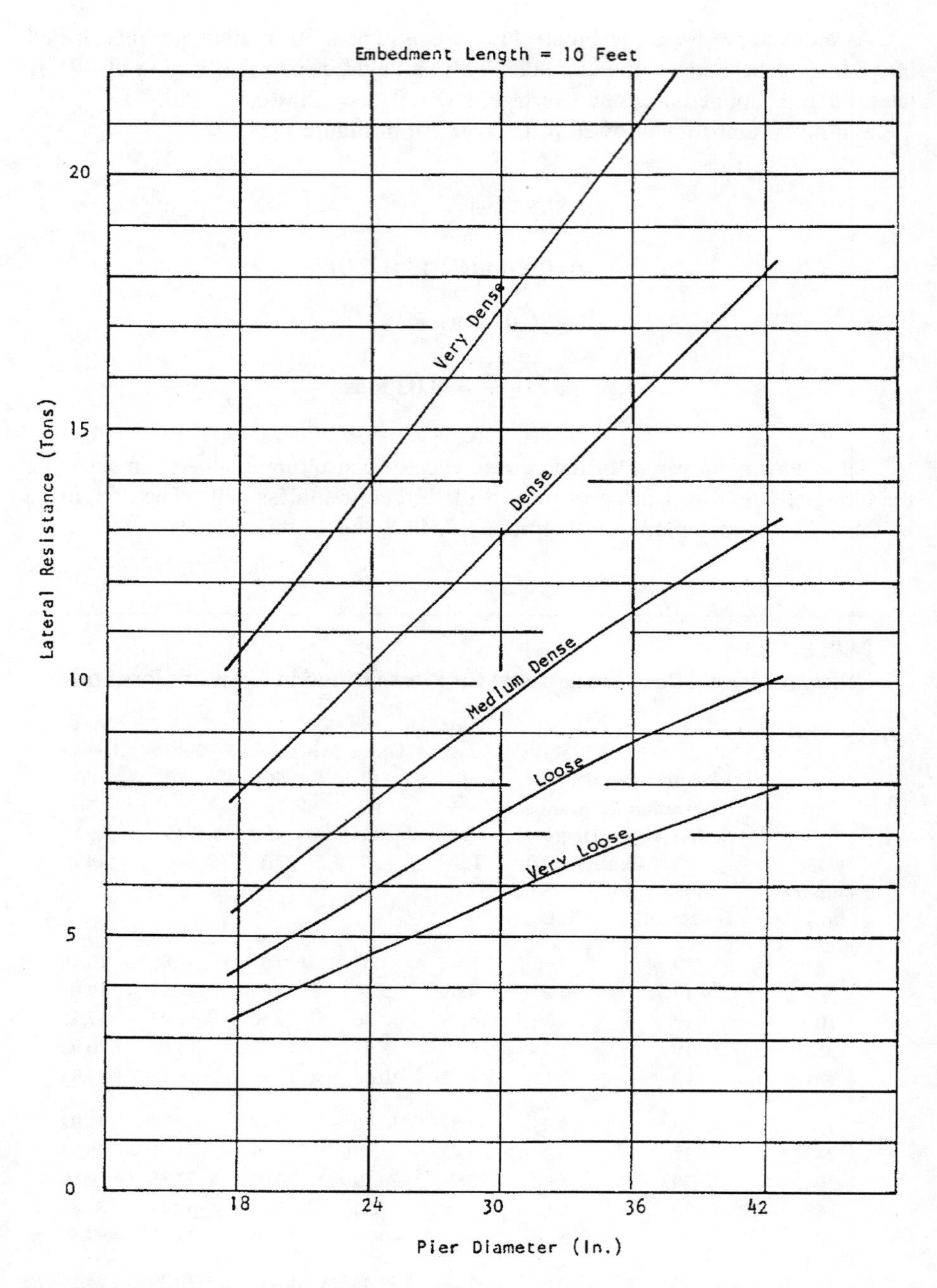

FIGURE 11.7 Lateral resistance of free-headed piers of various diameters drilled in cohesionless soils.

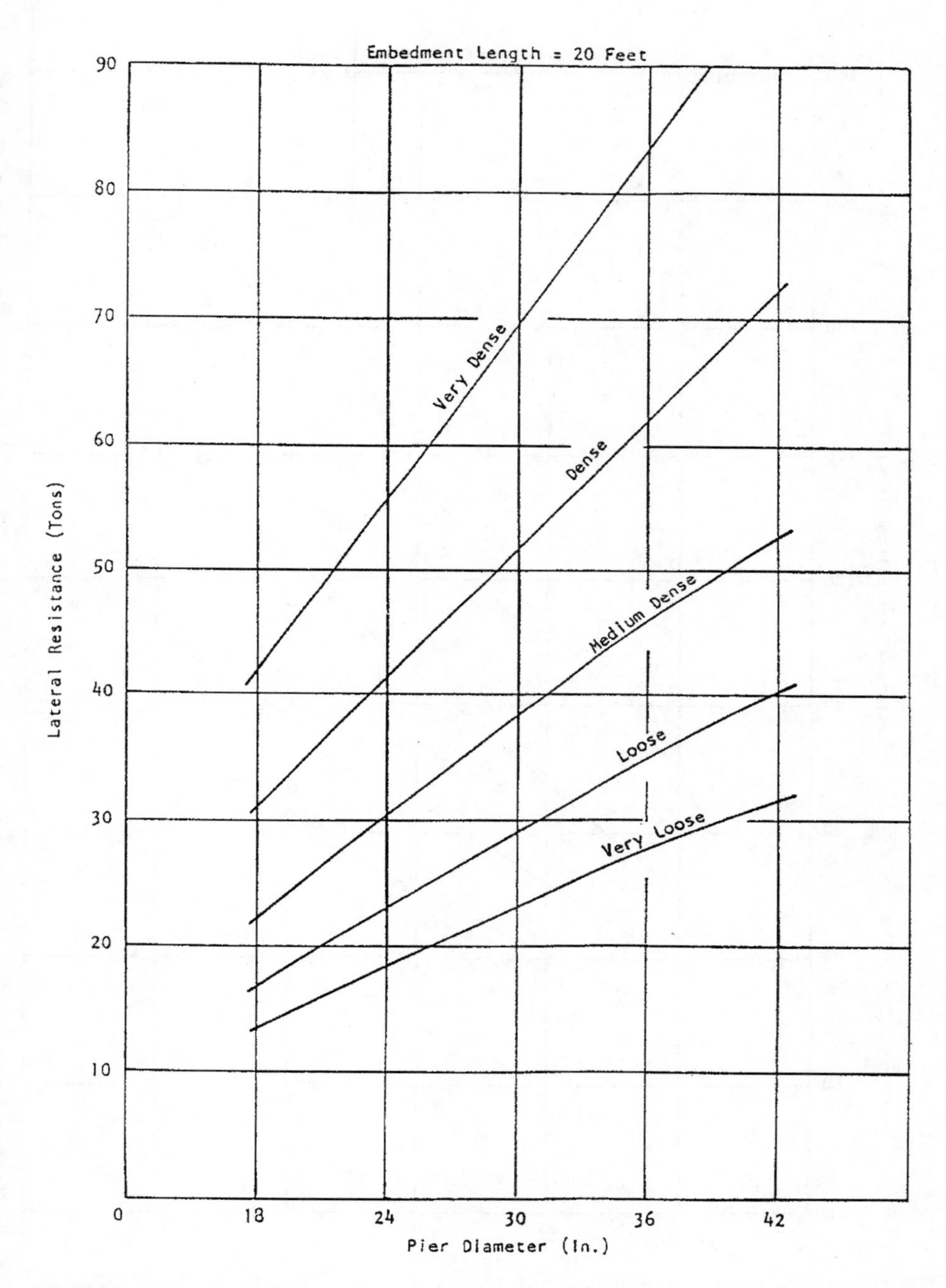

FIGURE 11.8 Lateral resistance of free-headed piers of various diameters drilled in cohesionless soils.

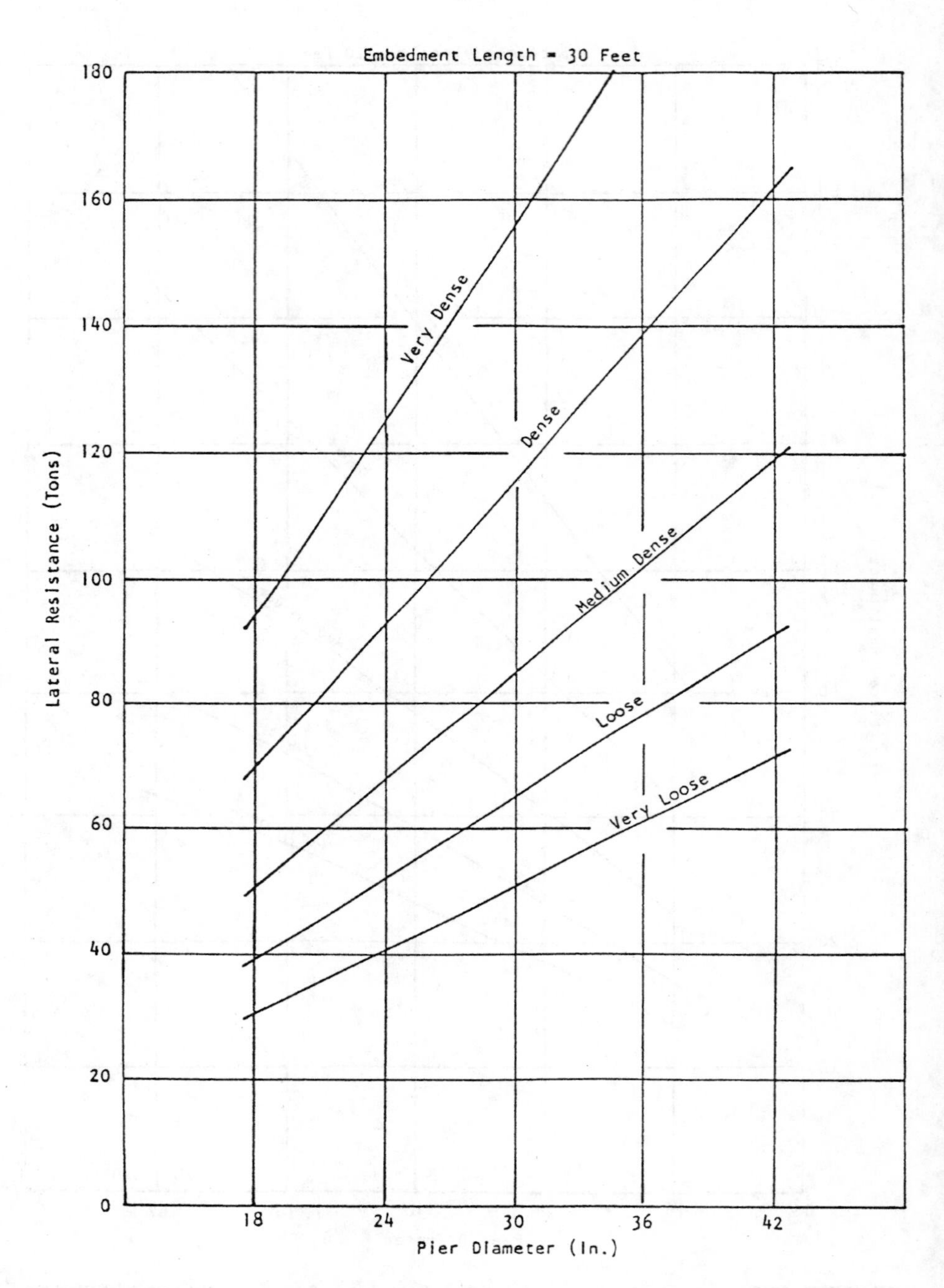

FIGURE 11.9 Lateral resistance of free-headed piers of various diameters drilled in cohesionless soils.

11.5 WORKING LOAD OF DRILLED PIERS ON COHESIVE SOILS

The working load of drilled piers in cohesive soils is associated with the subgrade reaction and stiffness of clays.

The term subgrade reaction indicates the pressure p per unit of area of the surface of contact between the loaded beam and the subgrade on which it transfers loads. The coefficient of subgrade reaction k_s is the ratio between the pressure at any point of the surface of contact and the settlement y produced by the load application at that point.

$$k_s = p/y$$

The value k_s depends on the elastic properties of the subgrade and on the dimensions of the area acted upon by the subgrade reaction. The value is expressed as either lbs/in^3 or tons/ft^3.

Figure 11.10 represents a vertical pier with width D, which is drilled into the subgrade. Before any horizontal forces have been applied to the pier, the surface of contact between the pier and subgrade are acted upon at any depth z below the surface by pressure p_o, which is equal to the earth pressure at rest.

If the pier is moved to the right, the pressure on the left-hand face drops to a very small value. At the same time, the pressure p_p on the right-hand face increases from its initial value p_o to a value p'_o that is somewhat greater than the earth pressure at rest $p_{o\cdot}$. Hence, at the outset of the movement of the pier toward the right, $y_l = 0$ and the pressure on the two faces of the piers is at any depth z.

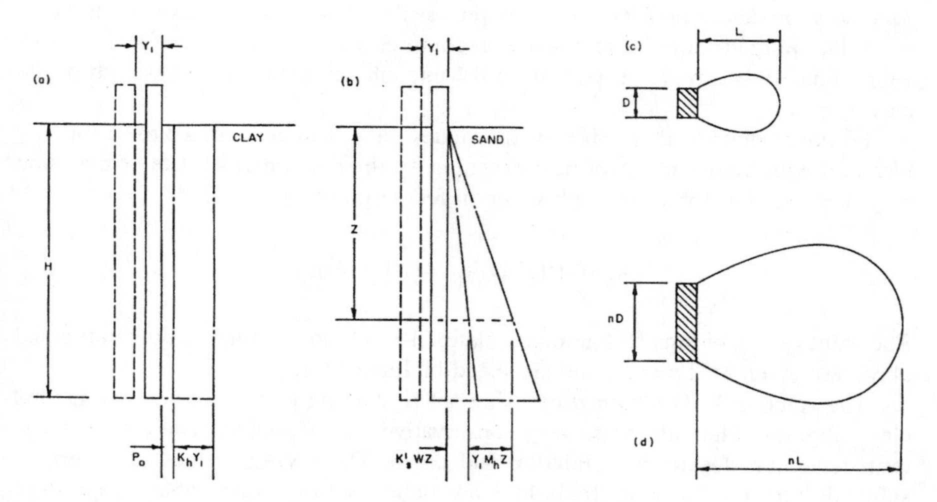

FIGURE 11.10 Vertical beam embedded (a) in stiff clay and (b) in sand; (c) influence of width of beam on dimensions of bulb of pressure (after Broms).

$$p_a = 0 \text{ (left-hand side)}$$

$$p_p = p'_o + p$$

$$= p'_o + k_h\, y_l \text{ (right-hand side)}$$

where $p = k_n\, y_l$ is the increase of the pressure on the right-hand face due to the displacement y_l of the pier.

The deformation characteristics of stiff clays are more or less independent of depth. Therefore, at any time, the subgrade reaction p is almost uniformly distributed over the right-hand face of the pier and the coefficient of the subgrade reaction is:

$$k_n = p/y_l$$

Because of progressive consolidation of clay under a constant load, the value y_i increases and the value k_h decreases with time. In clay as well as in sand, the horizontal displacement y increases in simple proportion to the width D, $y_n = n\, y_l$.

For a beam surrounded by clay,

$$k_n = 1/D\ k_{hl}$$

where k_{hl} is the coefficient of horizontal subgrade reaction for a vertical beam with a width of 1 ft embedded in clay.

As a basis for estimating the coefficient of subgrade reaction k_s, the value k_{s1} for a square plate with a width of 1 ft has been selected, because this value can, if necessary, be determined by averaging the results of several load tests in the field.

If the subgrade consists of heavily precompressed clay, the value of k_{sl} increases approximately in simple proportion to the unconfined compressive strength of the clay q_u.

For piers embedded in stiff clay, the values for k_{hl} can be assumed to be roughly identical with the value k_{sl} for beams resting on the horizontal surface of the same clay. The value k_h for a pier with width in feet is given by

$$k_h = (1/1.5\ D)k_{hl} = (1/1.5\ D)k_{sl}$$

The values of k_{sl} in tons/ft^3 for square plates, 1 × 1 ft and resting on precompressed clays, are given by Terzaghi and are listed in Table 11.3.

The value of k_{sl} is obtained by using a 1 ft^2 bearing plate on a preconsolidated clay subgrade. The values are very conservative and undoubtedly allow for long-term reduction due to consolidation and creep. The revised values of a vertical subgrade reaction listed in Table 11.3 are believed to be reasonable and a more realistic approach.

If the subgrade consists of preconsolidated clay, the coefficient of horizontal subgrade reaction has the same value k_h for every point of the surface of contact.

TABLE 11.3
Coefficient of Subgrade Reaction

Consistency of clay	Stiff	Very Stiff	Hard
Value of q_u (ton/ft^2)	1–2	2–4	over 4
Range of k_{sl} square plate	50–100	100–200	over 200
Proposed values	75	150	300

The maximum working load for a free-headed short pier can be calculated by the following equation

$$p = \frac{y\,k_h\,DL}{4[1 + 1.5\,e/L]}$$

When e/L = 0

$$p = \frac{y\,k_h\,DL}{4}$$

Table 11.4 is prepared on the basis of the above equation. The following should be observed:

1. All calculations are based on short, rigid conditions and no consideration has been given to long, elastic piers. As shown in Table 11.4, the stiffness factor indicates that the calculation for long pier behavior is unnecessary.
2. Under a lateral pressure, it is assumed that the pier rotates as a unit around a point located some depth below the ground surface.
3. The working loads listed in Table 11.6 are on the basis of 0.5 in. deflection, which is believed to be reasonable for a free-headed condition.
4. From Table 11.5, the lateral soil pressure for various consistencies of clay expressed as lb/ft^2 of depth is as shown. Also listed is the recommended value given by Czerniak.

The conventional value of lateral soil pressure is very conservative, especially in the case where the piers are embedded in bedrock.

A restrained pier with a βL value less than 0.5 behaves as an infinitely stiff pier and the lateral deflection at the ground surface can be calculated directly from the equation:

$$p = y\,k\,D\,L$$

It should be noted that an increase of pier length appreciably decreases the lateral deflection at the ground surface for short piers. However, a change of pier stiffness has only a small effect on the lateral deflection for such piers. The lateral deflection

TABLE 11.4
Values of Stiffness Factors for Piers of Various Embedment Length and Diameter Drilled in Cohesive Soils of Various Consistencies

	Consistency	Soft	Medium Stiff	Stiff	Medium Hard	Hard
	K_hD (tons/ft³)	1.67	33.3	66.7	133.3	266.6
Pier Diameter (in.)	K_hD (lbs./in.³)	19.3	38.5	77.0	154.2	307.9
	Pier Length (ft)					
18	120	0.51	0.60	0.72	0.85	1.01
24	120	0.38	0.45	0.53	0.63	0.75
30	120	0.30	0.36	0.42	0.51	0.60
36	120	0.25	0.30	0.35	0.42	0.50
42	120	0.21	0.25	0.30	0.36	0.43
18	240	1.02	1.20	1.44	1.70	2.02
24	240	0.76	0.90	1.06	1.26	1.50
30	240	0.60	0.72	0.84	1.02	1.20
36	240	0.50	0.60	0.70	0.84	1.00
42	240	0.42	0.50	0.60	0.72	0.86
18	360	1.53	1.80	2.16	(2.55)	(3.03)
24	360	1.14	1.35	1.59	1.89	(2.25)
30	360	0.90	1.08	1.26	1.53	1.80
36	360	0.75	0.90	1.05	1.26	1.50
42	360	0.63	0.75	0.90	1.08	1.29

0.51 indicates short pier (βL less than 2.25)
(2.55) indicates long pier (βL greater than 2.25)

TABLE 11.5
Lateral Soil Pressure for Various Consistencies

Consistency	Soft	Medium Stiff	Hard
Table Value	346	692	2768
Czerniak	100	300	400

at the ground surface of short, fixed piers is theoretically one fourth of those for the corresponding free-headed piers.

TABLE 11.6
Maximum Working Load (tons) on Free-Headed Piers of Various Diameters and Lengths Drilled in Cohesive Soils of Various Consistencies with Lateral Deflection of 0.5 in.

Pier Diameter (in.)	State / Vertical Subgrade Reaction K_{sl} / K_hD / Pier Length (ft)	Soft	Medium Stiff	Stiff	Medium Hard	Hard
	Vertical Subgrade Reaction K_{sl}	25	50	100	200	400
	K_hD	16.7	33.3	66.7	133.3	266.6
18	10	1.73	3.46	6.92	13.84	27.68
24	10	1.73	3.46	6.92	13.84	27.68
30	10	1.73	3.46	6.92	13.84	27.68
36	10	1.73	3.46	6.92	13.84	27.68
42	10	1.73	3.46	6.92	13.84	27.68
18	20	3.46	6.92	13.84	27.68	55.36
24	20	3.46	6.92	13.84	27.68	55.36
30	20	3.46	6.92	13.84	27.68	55.36
36	20	3.46	6.92	13.84	27.68	55.36
42	20	3.46	6.92	13.84	27.68	55.36
18	30	5.19	10.83	20.76	(41.52)	(83.04)
24	30	5.19	10.83	20.76	41.52	(83.04)
30	30	5.19	10.83	20.76	41.52	83.04
36	30	5.19	10.83	20.76	41.52	83.04
42	30	5.19	10.83	20.76	41.52	83.04

Notes: K_{sl} = Vertical Subgrade Reaction (ton/ft³)
K_h = Horizontal Subgrade Reaction (ton/ft³)
$K_hD = K_{sl}/1.5$
(83.04) indicates long pier behavior

11.6 WORKING LOAD OF DRILLED PIERS ON COHESIONLESS SOILS

If the subgrade consists of cohesionless soils, the value of the coefficient of horizontal subgrade reaction is determined by:

$$k_h = m_h z$$

and the value of m_h is assumed to be the same for every point of surface of contact. In cohesionless soils, the values y_i and k_h are practically independent of time. However, the modules of elasticity of the sands increase approximately in simple proportion to the depth. Therefore, it can be assumed that the pressure p required to produce a given horizontal displacement y_l increases in direct proportion to depth z, as shown in Figure 11.10.

$$k_h = p/y_l = m_h\, z$$

If the pier is embedded in sand,

$$k_{hn} = m_{hn} = p/y_n = p/n\, y_l$$

since $p/y_l = m_{hl}\, z$
$k_{hn} = (1/n)\, m_{hl}\, z = m_{hl} = m_{hl}\, D_l\, (z/n\, D_l)$

Substituting $k_{hn} = k_h$, $D_i = 1$ ft. $n\, D_l = D$ and $m_{hl}\, D_l = n_h$
$k_h = n_h\, (z/D)$

where n_h (tons/ft^3) is the constant of horizontal subgrade reaction for piers embedded in sands. It is seen that the value of k_h is assumed to approximately increase linearly with depth and to decrease linearly with the increasing width D of the loaded area. The coefficient of subgrade reaction is independent of the stiffness of the piers and the pier length.

The value n_h can be estimated by

$$n_h = A_w/1.35$$

in which w = the effective unit of sand
A = coefficient which only depends on the density of the sands. Ranges from 100 for very loose sand to 2000 for dense sands

From the above reasoning, Terzaghi prepared Table 11.7 for the evaluation at the constant of horizontal subgrade reaction n_h..

The stiffness factor for piers with various embedment lengths and diameters drilled in cohesionless soils with various relative densities is given in Table 11.8.

A laterally loaded pier behaves as an infinitely long member when μL is less than about 2.0 and as an infinitely long member when μL exceeds about 4.0. For short piers, an increase of the embedment length decreases the lateral deflections at the ground surface, while for long piers the lateral deflections at the ground surface are unaffected by the change of embedment length. On the other hand, an increase

TABLE 11.7
Coefficient of Horizontal Subgrade Reaction

	Coefficient n_h (tons/ft^3)		
Relative Density	**Loose**	**Medium**	**Dense**
Above Water	7	21	56
Below Water	4	14	34

TABLE 11.8
Value of Stiffness Factor (μL) for Piers of Various Embedment Lengths and Diameters Drilled in Cohesionless Soils of Various Relative Densities

Pier Diameter (in.)	State n_h (tons/ft³) n_h (lbs./in.³) Pier Length (in.)	Very Loose 7 8.10	Loose 21 24.30	Medium 56 64.81	Dense 74 85.65	Very Dense 92 106.48
18	120	1.73	2.16	(2.52)	(2.70)	(2.82)
24	120	1.33	1.60	1.98	(2.11)	(2.22)
30	120	1.10	1.37	1.68	1.77	1.86
36	120	0.96	1.19	1.44	1.52	1.60
42	120	0.85	1.05	1.27	1.35	1.41
18	240	(3.46)	4.32*	5.04*	5.40*	5.64*
24	240	(2.66)	(3.20)	(3.96)	4.22*	4.44*
30	240	(2.20)	(2.74)	(3.36)	(3.54)	(3.72)
36	240	1.92	(2.38)	(2.88)	(3.04)	(3.22)
42	240	1.60	(2.10)	(2.54)	(2.70)	(2.82)
18	360	5.19*	6.48*	7.56*	8.10*	8.46*
24	360	(3.99)	4.95*	5.94*	6.33*	6.66*
30	360	(3.30)	4.11*	5.04*	5.31*	5.58*
36	360	(2.88)	(3.57)	4.32*	4.56*	4.80*
42	360	(2.55)	(3.15)	(3.81)	4.05*	4.23*

1.73 indicates short pier (μL less than 2.0)
5.19* indicates long pier (μL larger than 4.0)
(3.46) indicates μL between 2.0 and 4.0

in the stiffness EL of the pier section decreases the lateral deflection of long piers, but does not affect the lateral deflections of short piers.

For long piers, the lateral deflection y at the ground surface can be calculated directly from

$$y = [2.40\ p]/nh^{3/5}\ EL^{2/5}$$

For short piers, the lateral defection y at the ground surface for a free-headed condition can be calculated from

$$y = \frac{18\ p\left(1 + 1.33\frac{e}{L}\right)}{L^2\ n_h},$$

For e = 0, $y = \dfrac{18\ p}{L^2\ n_h}$

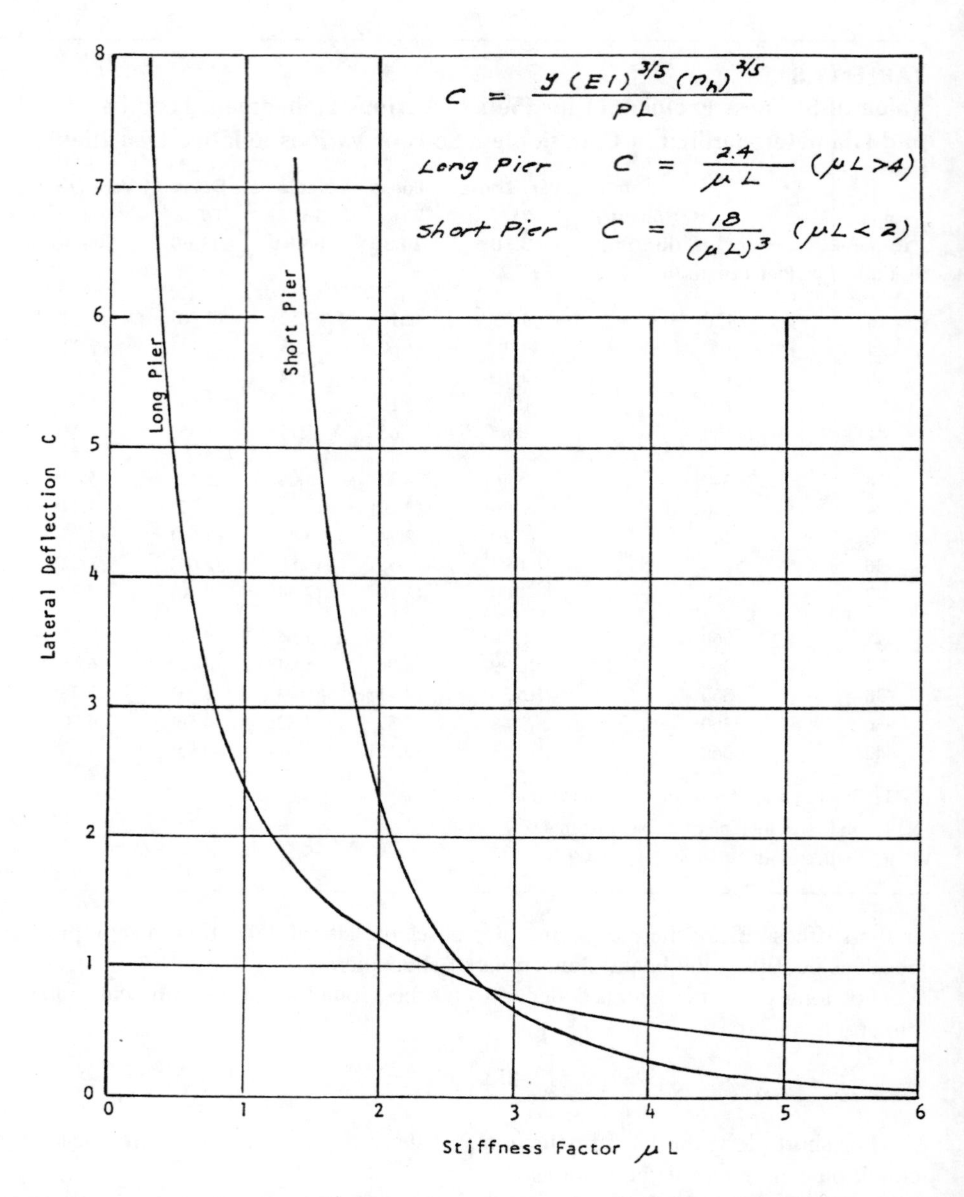

FIGURE 11.11 Lateral deflection related to stiffness factor for piers drilled in cohesionless soils.

The above equations can be plotted as shown in Figure 11.11. Table 11.9 is prepared on the basis of Figure 11.11, considering the cases of both long and short piers.

TABLE 11.9
Maximum Working Load (tons) for Free-Headed Piers of Various Diameters and Lengths Drilled in Cohesionless Soils of Various Relative Densities with Lateral Deflection of 0.5 in.

	State	**Very Loose**	**Loose**	**Medium**	**Dense**	**Very Dense**
	n_h (tons/ft³)	**7**	**21**	**56**	**74**	**92**
Pier Diameter (in.)	**Relative Density (%)** **Pier Length (in.)**	**20**	**30**	**50**	**70**	**80**
18	120	1.62	4.86	12.96 (17.43)	17.12 (17.30)	21.29 (20.10)
24	120	1.62	4.86	12.96	17.12	21.29 (32.14)
30	120	1.62	4.86	12.96	17.12	21.29
36	120	1.62	4.86	12.96	17.12	21.29
42	120	1.62	4.86	12.96	17.12	21.29
18	240	6.48 (4.26)	(8.28)	(14.86)	(17.30)	(20.10)
24	240	6.48 (6.81)	19.44 (13.24)	51.84 (23.75)	(27.65)	(32.12)
30	240	6.48 (9.81)	19.44 (19.12)	51.84 (34.29)	68.48 (39.92)	85.16 (46.38)
36	240	6.48	19.44 (25.50)	51.84 (45.72)	68.48 (53.23)	85.16 (61.84)
42	240	6.48	19.44 (32.72)	51.84 (58.67)	68.48 (68.01)	85.16 (79.36)
18	360	14.59 (4.26)	(8.28)	(14.86)	(17.30)	(20.10)
24	360	14.59 (6.81)	(13.24)	(23.75)	(27.65)	(32.12)
30	360	14.59 (9.84)	(19.12)	(34.29)	(39.92)	(46.38)
36	360	14.59 (13.11)	43.78 (25.50)	(45.72)	(53.23)	(61.84)
42	360	14.59 (16.82)	43.78 (32.72)	116.76 (58.67)	(68.01)	(79.36)

1.62 indicates lateral pressure on the basis of short pier
(4.26) indicates lateral pressure on the basis of long pier

11.7 PRESSUREMETER TEST

The value of horizontal subgrade reaction k_h can also be determined experimentally by using the Menard pressuremeter. The pressuremeter probe is inserted in the test hole at the desired depth. The radial expansion of the hole is expressed as a function of increasing radial pressures applied to its wall, similar to a common load test. Thus, the deformation modulus E_s can be determined at any depth. The following formula is used in the determination of the coefficient of horizontal subgrade reaction:

$$\frac{1}{kh} = \frac{1+\mu}{3\ Es} Ro \left[\frac{R/2}{R_o} C_2 \right]^{C_1} + \frac{C_1}{4.5\ Es} C_3 \frac{D}{2}$$

In which

k = Coefficient of horizontal subgrade reaction (kg/cm^3)
μ = Poisson's ratio (0.3 for most soils)
R_o = Radius of pressuremeter probe
D = Pier diameter (cm)
C_1 = Structural coefficient of soil (0.33 for sand, 0.66 for claystone)
C_2 = Shape factor in shear deformation zone (2.65 where L/D less than 20)
C_3 = Shape factor in consolidation zone (1.50 where L/D less than 20)
E_s = Deformation modules (kg/cm^2)

As an example, an actual pressuremeter test was made in a medium dense sand and gravel at a depth 16 ft below the surface. E_s value obtained from the test was 115 kg/cm^2. For a 42-in. diameter pier, the k_h value is:

$$1/K_h = [(1+0.33)/(3 \times 115)](30)[(53/30) \times 2.65](0.33)$$

$$+[0.33/(4.5 \times 115)] \times 1.5 \times 53$$

$$= 0.192 + 0.0507$$

$$= 0.2427$$

$$K_h = 4.12\ kg/cm^3 = 128\ ton/ft^3$$

By substitution, the horizontal subgrade reaction is

$$n_h = (128 \times 3.5)/15 = 29.9\ ton/ft^3$$

This value is reasonable in comparison with Terzaghi's value listed in Table 11.3.

11.8 APPLICATIONS

This chapter summarizes past research about lateral load on piers. The content is quoted directly from Broms' papers without alteration. It is intended that a rational procedure in designing piers for lateral load be established (Figure 11.12). The conventional building code lateral load values assigned to the piers or piles are

FIGURE 11.12 Microwave tower on piers subjected to lateral load.

undoubtedly low and the soil classification ambiguous. The following is a summary of the above design procedure.

1. Determine the predominant soil conditions surrounding the piers. It would be sufficient to indicate that the soil consists of, for instance, 15 ft of granular soils and 5 ft of claystone bedrock. Refinement such as the existence of clay lenses and cobbles will not be necessary.
2. In the case of granular soils, select the loose state as design criteria. Direct shear tests and density tests should be conducted, and the average value selected for the design purpose.
3. It is important to establish the ground water level, so that the appropriate constant of subgrade reaction value can be selected.
4. In the case of cohesive soil, the unconfined compressive strength tests should be conducted on many samples and the average value used for design. More important is the fact that the penetration resistance value should not be overlooked.
5. In the case of claystone bedrock, the unconfined compressive strength is usually only a fraction of its actual strength; consequently, more attention should be directed to the penetration resistance values.
6. The most direct and accurate method of determining the soil strength is by use of the Menard pressuremeter. Pressuremeter tests can be conducted at relatively low cost, and the results are more reliable than laboratory testing on samples in which a great deal of disturbance is to be expected.
7. For the design of structures such as sign posts and transmission towers, where large deflection can be tolerated, lateral soil resistance given in Tables 11.1 and 11.2 should be used.

8. For high-rise structures subject to wind load and seismic load, or high retaining walls subject to earth pressure, lateral deflection can be critical. In such cases, a determination of long or short piers' conditions should be made. For low-rise buildings, a restrained condition can generally be assumed. If a free-headed condition is selected, the tolerable deflection of 0.5 in. can be assumed. Tables 11.6 and 11.9 can be used as guides.

REFERENCES

B.R. Broms, Lateral Resistance of Piles in Cohesionless Soils, *Journal of the Soil Mechanics and Foundation Division*, ASCE, Vol. 90, No. SM3, proc. paper 3909, 1964.

B.R. Broms, Lateral Resistance of Piles in Cohesive Soils, *Journal of the Soil Mechanics and Foundation Division*, ASCE, Vol. 90, No. SM2, Proc. Paper 3825, 1965.

B.R. Broms, Design of Laterally Loaded Piles, *Journal of the Soil Mechanics and Foundation Division*, ASCE, Vol. 91, No. SM3, Proc. paper 4342, 1965.

E. Czerniak, Resistance to Overturning of Single, Short Pile, *Journal of the Structural Division*, ASCE, Vol. 83, No. ST2, proc. paper 1188, 1957.

P. Kocsis, *Lateral Loads on Piles,* Bureau of Engineering, Chicago, IL, 1968.

H. Matlock and L.C. Reese, Generalized solutions for Laterallly Loaded Piles, *Journal of the Soil Mechanics and Foundation Division*, ASCE, Vol. 86, No. SHS, proc. paper 2626, 1960.

K. Terzaghi, Evaluation of Coefficient of Subgrade Reaction. *Geotechnique,* London, Vol. V., No. 4, 1955.

R. Woodward, W. Gardner, and D. Greer, *Drilled Pier Foundations,* McGraw-Hill, New York, 1972.

12 Driven Pile Foundations

CONTENTS

With exception of footings, probably the oldest foundation system is the driven pile foundation. Wooden piles were driven by stone hammers, hauled up by use of pulleys, and dropped from a platform by gravitational force. Many historical structures were founded on piles driven through soft soils into firm bearing strata.

The function of a pile foundation is essentially the same as a pier foundation, as discussed in previous chapters. The major differences between the uses of a pile and a pier foundation are:

1. The diameter of an individual pile as well as its load-carrying capacity is limited.
2. Large diameter piers are used to support high column loads, while a pile group is used for the same purpose.
3. Pile driving technique and pier installation procedures are different. Both require special equipment and specialized contractors.

The analysis on the lateral pressure against the piers, as described in Chapter 11, is also applicable in the case of piles. However, a single pile is seldom used.

5. Defects of a driven pile cannot be easily detected, while a pier shaft can be inspected by entering the hole.
6. Piers are widely used in expansive soil areas to prevent heaving. The use of piles for such a purpose is still under study.

The selection of the use of a pile or pier foundation system depends on the type of structure, the regional subsoil conditions, the water table level, the available equipment, and many other factors.

The types of pile commonly used are as follows:

Timber piles — For centuries, timber piles have been used to support structures founded on soft ground. The entire city of Venice was founded with timber piles over the muddy deposit on the River Po. An individual pile is limited in diameter as well as length. The length is generally limited to around 60 ft. Timber piles can be damaged by excessive driving and by decay. Today, commercial piles are usually treated by chemicals that prevent decay and increase their life.

Concrete piles — Precast concrete piles may be made in various shapes and diameters. Reinforced precast concrete piles are sometimes prestressed to ease driving and handling. The length of concrete piles is limited to the capability of handling equipment. To increase the length limitation, consideration has been given to the possibility of splicing the piles. Cast-in-place piles are similar to piers, but not as flexible in capacity. Concrete piles are generally not susceptible to deterioration.

A great deal of publicity has been launched by various companies to increase the market use of concrete piles. Raymond piles are widely used in Asian countries where adequate timbers are scarce.

Steel piles — Steel piles are usually either pipe-shaped or H-sections. Pipe-shaped steel piles may be filled with concrete after being driven. H-shaped steel piles can be driven to a great depth through stiff soil layers and will not easily be deflected when encountering cobbles. Steel piles are subjected to corrosion. In strong acid soils such as fill or organic matter, and in sea water, corrosion is more serious.

Composite piles — Composite piles are a combination of a steel or timber lower section with a cast-in-place concrete upper section. The uncased Franki concrete pile is formed by ramming a charge of dry concrete in the bottom of a steel casing so that the concrete grips the walls of the pipe and forms a plug. A hammer falling inside the casing forces the plug into the soil, dragging the casing downward by friction. At the bearing level, the casing is anchored to the driving rig, and the concrete plug is driven out its bottom to form a bulb over 3 ft in diameter. The casing is then raised while successive chargers of concrete are rammed in place to form a rough shaft above the pedestal. Franki piles are widely used in Hong Kong, where the subsoil consists of alternate layers of soft soil and hard rocks. Most high-rise structures in Hong Kong are founded with Franki piles.

12.1 ALLOWABLE LOAD ON PILES IN COHESIONLESS SOILS

Piles in cohesionless soils derive their load-carrying capacity from both point resistance and friction on the shaft. The relative contributions of point resistance and shaft resistance to the total load-carrying capacity of the pile depend on the density and shear strength of the soil and on the characteristics of the piles. An empirical method in determining the load-carrying capacity of the pile is based on the results of the standard penetration test. A more exact method is based on the theory of plasticity.

12.1.1 Penetration Test Method

A simple and direct method in the determination of bearing capacity of piles driven in cohesionless soils is by the utilization of the results of the standard penetration resistance. Since such values are obtained in all field investigations, no additional tests will be required.

$$Q_f = 4N\, A_p + \frac{N_a\, A_s}{50}$$

where Q_f = ultimate pile load, tons
N = Standard penetration resistance at pile tip, blows per ft
A_p = cross-sectional area of pile tip, in ft^2
N_a = average penetration resistance along the pile shaft, blows per ft
A_s = surface area of the pile shaft, ft^2

Since this is an empirical method, a factor of safety of at least three is used. Therefore, the allowable load Q_a is determined as follows

$$Q_a \le Q_f/3$$

For non-displacement piles such as H-piles, a factor of safety of four is recommended.

For cone penetration resistance value, the ultimate bearing resistance value is suggested as follows:

$$Q_a = q_p\, A_p$$

where q_p is the average cone penetration resistance with a limiting value of 15 MN/m^2.

12.1.2 Plasticity Method

Large-scale experiments and measurements on full-scale piles have shown that the skin friction per unit of area does not increase with depth below a critical depth (Hc), which for all practical purposes is equal to:

$$H_c = 20D$$

where D is the diameter or width of the pile.

For piles with a length in granular soil less than the critical depth (H_c), the ultimate point resistance is given by:

$$q_p = \gamma\ L\ N_q$$

where q_p = ultimate point resistance, lb/ft^2
γ = effective unit weight of the soil, lb/ft^3
L = length of pile embedment, ft
N_q = a bearing capacity coefficient

For piles with lengths in excess of the critical depth, the ultimate point of resistance is constant and equal to:

$$q_p = \gamma\ H_c\ N_q$$

Values of Berezontzev's factor N_q as plotted conveniently by Tomlinson are shown in Figure 12.1.

The ultimate skin friction acting on the pile of length L is related to the ultimate point of resistance by the equation

$$f_f = \frac{q_p}{\alpha}$$

where α = a coefficient related to the shearing resistance as shown in Table 12.1.

It is recommended that a factor of safety of three be applied to q_p and f_f. Hence, the allowable load (Q_a) on a pile in cohesionless soil is computed as follows:

For $L < H_c$

$$Q_a = \frac{1}{3}\left[q_p\ A_p + \frac{f_f}{2} A's\ L\right]$$

where q_p and f_f are computed at depth L

and A_p = cross-sectional area of pile tip ft^2
$A's$ = unit surface area of the pile shaft, ft^2/ft

For length of pile exceeding the critical length of 20 ft

$$Q_a = \frac{1}{3}\left[q_p\ A_p + \frac{f_f}{2} A's\ Hc + f_f\ A's(L - Hc)\right]$$

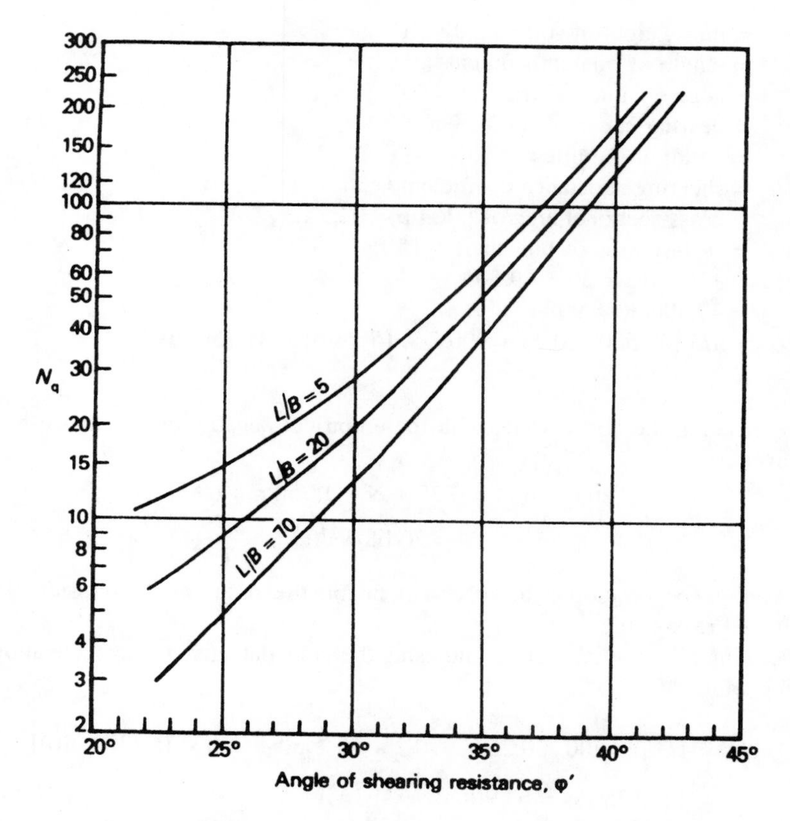

FIGURE 12.1 Berezontzeyv's bearing capacity factor (after Tomlinson).

TABLE 12.1
Friction Angle and Shearing Resistance
(α as a function of shearing resistance φ)

φ	25°	30°	35°	40°
soil density	loose	compact	dense	V dense
α	35	50	75	110

(after Vesic)

where q_p and f_f are computed at depth H_c = 20 D

12.1.3 Example

A 12-in.-square section of concrete pile is driven to an embedded depth of 15 ft in a cohesionless soil, which has the following properties:

γ = unit weight of soil = 110 lb/ft^3
L = length of pile embedment = 15 ft
ϕ = angle of internal friction = 30°
N_q = bearing capacity coefficient = 30
D = width of the pile = 1 ft
α = shearing resistance coefficient = 50
A_p = cross-sectional area of pile tip = 0.25 ft^2
$A's'$ = surface area of pile shaft = 15 ft^2
q_p = 110 × 15 × 30 = 49,500
f = 49,500/50 = 990
Q_a = 1/3 [49,500 × 0.25 + 990/2 × 15 × 15] = 41,250 lbs
= 20.6 tons

Upon checking the above value with the empirical penetration resistance method, we have:

$$20.6 = 4\text{ N} \times 0.25 + \text{N} \times 15/50 = 1.03\text{ N}$$

$$\text{N} = 20\text{ blows/ft}$$

The average penetration resistance will be on the order of 20 blows/ft, that is considered reasonable.

For long piles with L = 30 ft, and using the same data given above, the allowable bearing value is:

$$Q_s = 1/3\ [49{,}500 \times 0.25 + 990/2 \times 15 \times 30 + 990 \times 15\ (30 - 20)]$$

$$= 127{,}875\text{ lbs} = 63.9\text{ tons}$$

Upon checking with the empirical method, this corresponds to a penetration resistance of about 43 blows/ft. This value is high; it is doubtful that by increasing the pile embedment to 30 ft, the penetration resistance can be doubled. The deviation between the empirical method and the plasticity method becomes more pronounced as the length of embedment into the soil increases.

12.2 ALLOWABLE LOAD ON PILES IN COHESIVE SOIL

In contrast to a friction pile in sand, the point resistance of a pile embedded in soft clay is usually insignificant. It seldom exceeds 10% of the total capacity. Piles driven in cohesive soils generally derive their load-carrying capacity from friction or shaft adhesion. However, in very stiff clays large magnitudes of point resistance do develop and contribute to the total bearing capacity of the pile.

The shear strength or side resistance per unit of contact area depends largely on the properties of the clay and except for very long piles does not depend on the depth of penetration.

Various methods have been used to compute the distribution of the load imposed on the pile; no reliable results have been obtained. This is because homogeneous soils seldom exist in nature, and the stress deformation properties of the soil cannot be readily determined.

The bearing capacity of piles embedded in clay can be evaluated by using the total stress method or the effective stress method.

12.2.1 Total Stress Method

The state of stress around the pile is very complicated. It changes during and immediately after the driving. In spite of the influence of disturbance and other factors, the ultimate shear strength mobilized between the pile and the embedded clay is approximately equal to the undrained shear strength from the unconfined compressive strength test. It is assumed that most of the passive pressure developed at the tip of the pile will tend to dissipate with time and, therefore, the point resistance can be neglected. Hence, the ultimate bearing capacity of a pile embedded in soft clay can be estimated from this formula:

$$Q = \alpha_{z2}\, c_a\, c_d\, L$$

where Q = Ultimate load capacity, lb
α_z = Reduction factor
c_a = Adhesion factor
= $q_u/2$
q_u = Unconfined compressive strength, tons/ft^2
c_d = pile perimeter, ft
L = length of pile embedment, ft

The relationship between the reduction factor and the unconfined compressive strength of steel piles is shown in Figure 12.2.

Tomlinson determined the adhesion of piles embedded in soft clay, as shown in Table 12.2.

Using a factor of safety of three, the allowable pile capacity is computed as:

$$Q_a = \frac{Q_f}{3}$$

The shaft friction of piles driven in clay with an undrained shear strength in excess of approximately 2000 psf varies significantly, depending on the properties of the clay, time effects, driving methods, and pile types. In addition, piles in very stiff clay can exhibit substantial point resistance. Therefore, for piles in clay with an undrained shear strength in excess of 2000 psf, it is suggested that the ultimate bearing capacity be determined by pile load tests.

12.2.2 Effective Stress Method

Meyerhof and others in 1979 proposed that the skin friction per unit of area on a pile in clay should be expressed in terms of effective stress. The portion of a load carried by point resistance and skin friction can be calculated as follows:

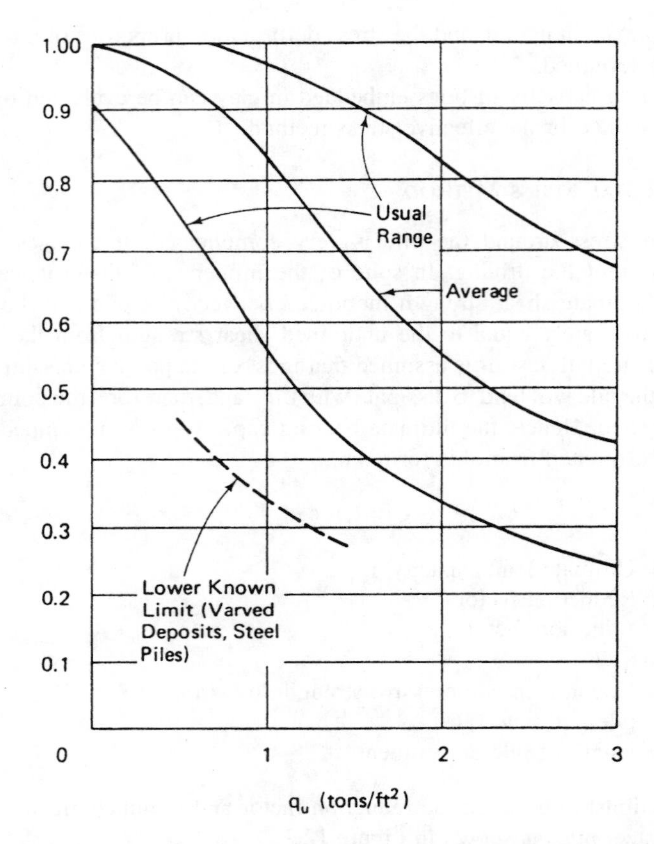

FIGURE 12.2 Reduction factor vs. unconfined compressive strength of friction piles in clays (after Peck).

TABLE 12.2
Adhesion of Piles in Saturated Clay

Material	Cohesion (psf)	Adhesion (psf)
Concrete	0–700	0–700
	750–1500	700–900
Timber	1500–2000	900–1300
Steel	0–750	0–600
	750–1500	600–750
	≥1500	75

$$Q_p = N_c C_u A_p$$

where Q_p = ultimate point load, lbs
N_c = bearing capacity factor (as a function of pile diameter, given as follows)
C_u = minimum undrained shear strength at the pile tip, lb/ft^2

Point diameter (inches)	N_c
<18	9
18–36	7
>36	6

$$Q_f = As\ \beta\ p_o$$

where Q_f = ultimate friction load, lb
A_s = surface area of the pile shaft, ft^2
p_o = effective overburden pressure at the considered depth, lb/ft^2
β = skin friction factor

On the assumption that the clay next to the pile is completely remolded, and thus that the cohesion intercept is zero and the effective angle is ϕ, then

$$\beta = K_s \tan \phi$$

where K_s = coefficient of earth pressure on the pile shaft
ϕ = effective angle of friction

Both K_s and ϕ are difficult to measure. However, available data indicates that for clay with undrained shear strength less than 2000 psf β varies from only 0.25 to 0.40 for piles less than 50 ft in length. Therefore, for design purposes, a typical value of 0.3 may be used. For lengths of piles in excess of 50 ft, the value of β should be altered according to this table, as indicated by Meyerhof:

Depth (ft)	β
0–50	0.30
50–100	0.25
100–150	0.20
>150	0.15

(after Meyerhof)

If the value of K_o (the coefficient of the earth pressure at rest) is not known for piles driven into stiff clays, the corresponding skin friction factor can only be estimated with a wide range. Some tests show that β varies from 0.5 to 2.5 for stiff clays.

The results of various investigations on this subject have shown some improvement in the prediction of allowable capacity for piles driven into stiff clays, yet they have not reached the stage of practical application. The effective stress method is included in this chapter only for information to the consultants.

12.2.3 Example

A 12-in. diameter concrete pile is driven in a normally consolidated clay to an embedded depth of 35 ft. The clay has the following properties:

γ = unit weight of soil = 110 lb/ft^3
q_u = unconfined compressive strength = 1400 psf
c_a = unit adhesion, with q_u = 0.7 ton/ft^2 = 0.9 × 700 = 630 psf
ϕ = angle of internal friction = 0
c = unit cohesion = 1400/2 = 700 psf
A = surface area of pile = πc_d L = 3.14 × 35 = 110 ft^2
c_d = pile perimeter, ft

To find the allowable bearing capacity with a factor of safety of two,

$$Q_f = c_a \, c_d \, L$$

$$= 630 \times 110$$

$$= 69{,}300 \text{ lbs.} = 34.65 \text{ tons}$$

$$Q_a = 34.6/2 = 17.325 \text{ tons}$$

12.3 PILE FORMULAS

The bearing capacity of a point-bearing pile may be approximately equal to the resistance of the soil against rapid penetration of the pile under the impact of the pile-driving hammer. Hence, many attempts have been made to develop "pile-driving formulas" by equating the energy delivered by the hammer to the work done by the pile as it penetrates a certain distance against a certain resistance, with allowance made for energy losses.

12.3.1 Engineering News Record Formula

As early as 1888, Wellington developed the Engineering News Formula as follows:

$$Q_{dy} = 2 \text{ W H/(S+C)}$$

where Q_{dy} = allowable pile capacity, lb
W = weight of hammer, lb
H = height of fall of hammer, ft
S = amount of pile penetration per blow, in./blow
C = 1.0 for drop hammer
C = 0.1 for steam hammer

The Engineering News Record Formula has a built-in factor of safety of six. Pile load tests have shown that the real factor of safety of the formula averages two instead of apparent six and that the safety factor can be as low as 2/3 and as high as 20.

Because of its simplicity, the formula has been widely used, but the results of load tests on driven piles have shown this formula is not really reliable for computing the load-carrying capacity. For consultants, the formula can at least be used as a rough guide. For small projects, where a load test is not justified and where over-design is permissible, such a formula can still be conveniently applied.

12.3.2 Danish Formula

Another pile-driving formula widely used in Europe is known as the Danish Formula. It is given by:

$$Q = e_h \; E_h / (2 + 1/2 \; S_o)$$

where Q = ultimate capacity of the pile
e_h = efficiency of pile hammer
E_h = manufacturers' hammer energy rating
S = average penetration of the pile from the last few driving blows
S_o = elastic compression of the pile

The Danish Formula is based on the assumption that the energy losses consist only of the elastic deformation of the pile and are moreover, not influenced by the penetration of the point of the pile. Statistical studies suggest that the formula should be used with a factor of safety of three.

12.3.3 Evaluation of Pile Formula

Numerous attempts have been made to take account of the energy losses in the pile formulas. Some of these attempts have resulted in very complicated expressions and procedures.

Pile-driving formulas are based on the assumptions that the bearing capacity of a driven pile is a direct function of the energy delivered to it during the last blows of the driving process, and that the energy transmission from the hammer to the pile and soil is instantaneous on impact.

These two assumptions have been proven wrong by many investigators. It has been clearly demonstrated that the bearing capacity of a pile is related, not so much to the total energy per blow of the driving system, but more importantly to the distribution of this energy with time at and after impact. From the many investigations of pile driving by means of the wave propagation theory, it has been made clear that the time effects as related to the propagation of impact forces on the pile have a governing influence on the behavior of piles during driving.

All pile-driving formulas involve the data of hammer blows. Hammer blow values are furnished by inspectors with sometimes little experience. As stated in Chapter 3, subsoil conditions vary a great deal even in very short distances. The blow count figure used in design can be off by as much as 100%.

The lack of reliability of pile-driving formulas was recognized a long time ago; for example, in 1942 Peck stated:

> It can be demonstrated by a purely statistical approach that the chances of guessing the bearing capacity of a pile are better than that of computing it by pile-driving formula…to determine the ultimate bearing capacity of a pile, the following procedure then would be justified: take 100 poker chips and label them with numbers so as to form a geometrically normal array having a mean value of 91 tons and a standard deviation of 1.55. Mix the poker chips and select one. The value written on the chip will be the bearing capacity of the pile. The value from the chip will be nearer to the true capacity more frequently than the value obtained by use of any of the pile formulas.

The lack of confidence that engineers have in such design formulas is demonstrated by the fact that the safety factors applied to determine the allowable loads are always very large; a value of F.S. = 6 is typical. This is a theoretical F.S. and not usually verified by full-scale load tests. Load test results have shown that the actual F.S. vary from less than 1 to values as high as 20.

Customary ranges of working load on driving piles are listed in Table 12.3. The values can be used for preliminary estimate in design.

TABLE 12.3
Ranges of Working Load on Driven Piles

Type	Load (kips)
Concrete, precast, or restressed	
10 in. diameter	4.0–13.0
15 in. diameter	15.0–45.0
Steel pipe, concrete filled	
1.0 × 0.2 in. pipe	0.7–1.2
1.6 × 0.4 in. pipe	22.5–26.9
Raymond Step-taper	0.9–15.6
Steel H-Section	
HP 10 × 42 in.	11.0–18.0
HP 14 × 117 in.	33.5–45.0

12.4 PILE GROUPS

In actual construction, the use of a single pile seldom or never exists. All structures are supported by pile groups. This is probably the major difference between the use of a pile and a pier. Theoretical analysis of the behavior of a single pile cannot be applied to a pile group.

Both theory and practice have shown that pile groups may fail as units by breaking in the ground before the load per pile becomes equal to a design safe load. Piles are usually placed in groups with center-to-center spacing typically between two and five shaft diameters. The pile heads are poured into a slab. The slab or the pile cap has the effect of constraining the group to act as a monolithic unit.

The minimum allowable pile spacing is often specified by local building codes. A building code may state that "The minimum center-to-center spacing of piles not driven to rock shall be not less than twice the average diameter of a round pile, nor less than 2 ft 6 in. For piles driven to rock, the minimum center-to-center spacing of piles shall be not less than twice the average diameter of a round pile, nor less than 1.75 times the diagonal dimension of a rectangular or rolled structural steel pile, nor less than two ft."

12.4.1 EFFICIENCY OF PILE GROUPS

The efficiency of a pile group is the capacity of a group of piles divided by the sum of the individual capacities of the piles making up the group. In the case where a group of piles is comprised of end-bearing piles resting on bedrock or on dense sand and gravel, an efficiency of 1.0 may be assumed. An efficiency of 1.0 can also be assumed where a group of piles is comprised of friction piles driven in cohesionless soils. In the case where a group of piles is composed of friction piles driven into cohesive soils, an efficiency of less than 1.0 is to be expected. A pile efficiency may be assumed to vary linearly from the value of 0.7, with a pile spacing of three times the diameter.

If the piles and the confined mass of soil are driven as a unit like a pier, the ultimate bearing capacity of the group can be given as follows:

$$Q_s = q_d\,BL + S\,s_u$$

where q_d = ultimate bearing, per unit of area
S = surface area of embeded portion of the pile group
= D(2B + 2 L)
B = breadth of pile cap
L = width of pile cap
s_u = average undrained shearing resistance of the soil between surface and bottom of the piles.

12.4.2 SETTLEMENT OF PILE GROUPS

The settlement of a pile group is larger than the settlement of a single pile and may be as much as five or ten times larger, depending on the size of the pile group. A

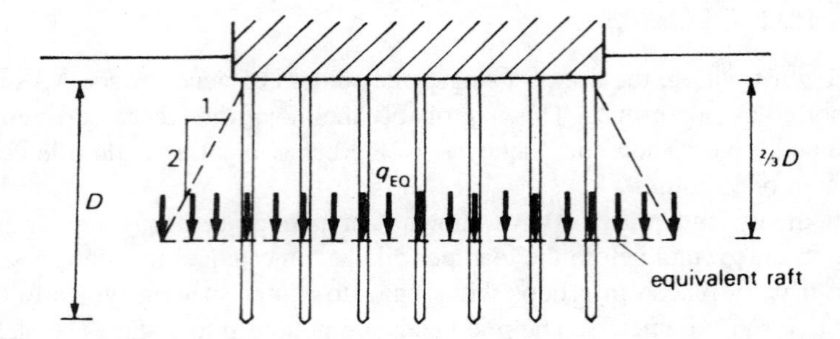

FIGURE 12.3 Settlement of pile group (after Whitlow).

pressure distribution analysis indicates that the larger the group, the deeper the stresses penetrate the strata, and thus, from the elasticity considerations, the larger the settlement. An individual pile has a small effective width; therefore, the stress influence from the pile load extends only to a shallow depth. With a pile group, the effective width increased dramatically; consequently, the stress influence is distributed to a much greater depth. As a result, a larger mass of soil is loaded, which in turn results in the consolidation of more material. The net result is a greater amount of settlement.

One way of estimating the settlement of a pile group is to consider an equivalent raft located at a depth of 2/3 D, where D is the embedded depth of the pile as shown in Figure 12.3.

The area of the equivalent raft is determined by assuming that the load spreads from the underside of the pile cap in the ratio of 1:2. The settlement due to the uniformly loaded equivalent raft is then calculated by conventional means with parameters obtained either from laboratory consolidation tests or from field penetration tests.

Because of the great numbers and diversity of the factors involved, it is very doubtful that the settlement of a pile group construction can be estimated within a practical range. At the present state of knowledge, it is preferable to consider every case individually and evaluate the probable settlement on the basis of the physical properties of the soil on which the load is transmitted by the piles.

12.5 NEGATIVE SKIN FRICTION

Negative skin friction is a force developed by friction between the pile and the surrounding soil. It is in a downward direction because of the movement of the soil mass; hence, the term downdrag.

Downdrag occurs when piles are driven through a layer of fill material that slowly consolidates due to its own weight; a downdrag is imposed on the pile shaft, resulting in pile settlement. A similar movement can take place when the pile is driven through a compressible material that is subject to overburden pressure, or when there is a sudden change of water level. This movement as a result of the friction force tends to carry the piles farther into the ground. The result could be excessive pile settlement.

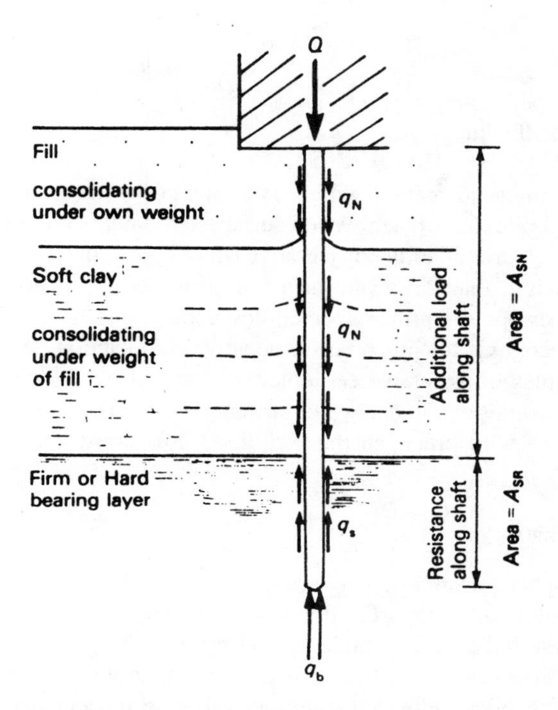

FIGURE 12.4 Negative skin friction (after Whitlow).

Because the soil alongside the pile is much more compressible than the pile itself, the downdrag forces on the upper part of the pile forces the lower part to slip downward with respect to the adjacent soils. Consequently, there is a neutral depth above which negative skin friction occurs and below which the skin friction acts upward. This is theoretically shown in Figure 12.4.

As consolidation proceeds, the magnitude of negative friction increases, since the effective overburden pressure increases as the excess pore pressure dissipates.

An estimate of the magnitude of downdrag can be obtained from an empirical analysis known as the β-method. The β-method utilizes the expression

$$f_s = \beta\ \sigma$$

where f_s = unit negative skin friction, ksf
σ = vertical effective stress
β = negative skin friction parameter
= 0.20 to 0.25 for clays
= 0.25 to 0.35 for silts
= 0.35 to 0.50 for sands

The total downdrag force, F_s, is contributed by each soil layer computed from the expression

$$F_s = f_s ph$$

where p = pile perimeter, ft
h = layer thickness, ft

This is an empirical method which is applicable when significant movement between soil and pile takes place. When surface settlements are less than 1 to 2 in., the estimated downdrag load is conservative. In addition, the method also overestimates the downdrag load on a pile in a group; therefore, in general, the method provides an upper bound on the expected downdrag loads.

Failure to recognize the potential for negative skin friction has been responsible for many examples of excessive settlement. As stated by Peck in 1996, "Indeed, recognizing the conditions under which it may develop is more important than the ability to estimate with precision the additional downward load or the amount of settlement."

12.5.1 Example

A steel building was originally supported by spread footings. Shallow wetting of the coarse granular soils supporting the footing resulted in over 3 in. of settlement. One corner of the building was underpinned by two 12 × 54 WF piles driven to a depth of 98 ft. Two years after the underpinning, deep wetting of the subsoil took place. One of the piles below the pile cap settled 8 in. and the adjacent ground subsided more than 12 in. A pull-out test performed on the failed pile indicated the skin friction force was at least 182 tons.

The building corner was again underpinned with three BP 12 × 74 piles driven to a depth of 145 ft. The upper 50 ft of each new pile was sleeved with an outer casing. The total design load for the column at this corner was about 240 kips.

The subsoil consisted of 35 ft granular soils, 58 ft of clayey soils, and 10 ft of silty gravel from a depth of 140 to 151 ft, as shown in the typical log in Figure 12.5. Free water was at depth of 101 ft. Figure 12.5 also indicates the change of moisture content of the subsoil in the past 3 years.

The upper dense granular soils compressed a small amount upon wetting and the lower fine granular soil had a collapse potential. If general wetting occurred, such as from poor surface drainage, settlement would generally be small, but continue for a long time. In contrast, the result of a point wetting source settlement, such as a water line break, will be large and sudden.

Driving criteria for the underpinning was to an ultimate bearing capacity of 60 tons, a factor of safety of two, and 100 tons allowed for negative skin friction. Negative skin friction value was based on an adhesion value of 1.0 ksf for the granular soil, 0.4 ksf for the clay, and a wetting depth of 70 ft.

About 1 year after the underpinning of the northwest corner was completed, excessive pile settlement again took place. Investigation by opening of test pits revealed that the underpinned pile had moved down at least 12 in. The soil adjacent to the pile was very wet. The source of wetting was from the overflow of water from the plant.

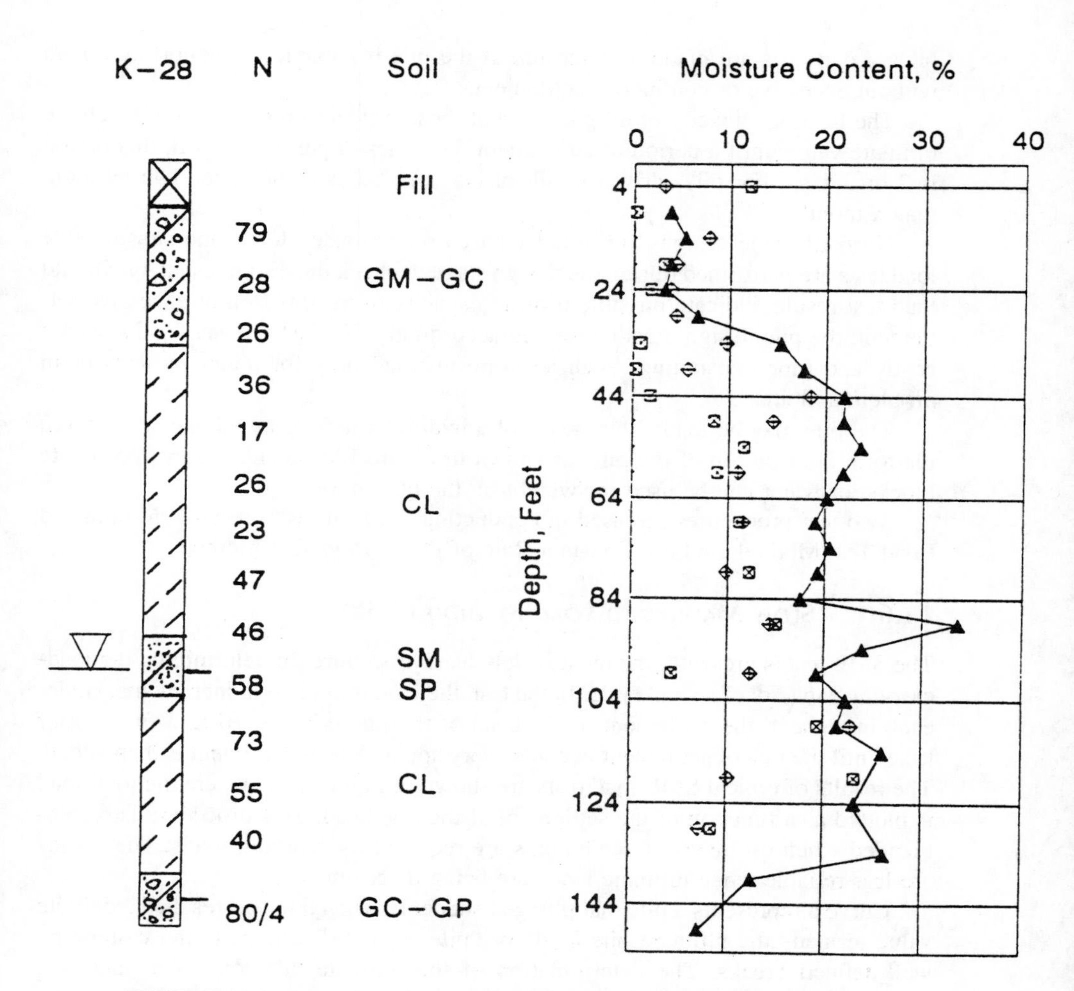

FIGURE 12.5 Log of test hole and moisture comparison at failure pile.

It was concluded that the wetting caused compression of the fine-grained layer that, in turn, caused the weight of all the soil and adjacent concrete pavement above to be carried by the piles. This increased load exceeded the bearing capacity of the one pile.

The pile failure and subsequent load test showed that negative skin friction is a significant force and can be easily underestimated. With the second underpinning complete, the performance of the building has been satisfactory.

12.6 PILE LOAD TESTS

The most reliable method of determining pile capacity is a load test. Pile load tests are made to determine the ultimate failure of a single pile or sometimes a group of piles. Load tests are made to determine if the pile is capable of supporting a load without excessive or continuous settlement.

The bearing capacity of all piles except those driven to rock does not reach the ultimate value until a period of adjustment. For piles in permeable soil, this period is 2 or 3 days. For piles driven in silt or clay, the adjustment period can be more than a month.

Normally, pile designs are initially based on estimated loads and subsoil. Pile load tests are performed during the design stage to check the design capacity. Should load test results indicate possible bearing capacity overestimation or excessive settlement, the pile design must be revised accordingly. Needless to say, load tests are costly and time consuming. Such tests are justified only for major projects or in problem soil areas.

The pile may be loaded by means of a hydraulic jack or by adding weight to a platform built on top of the pile, as shown in Figure 12.6. Sand, pig iron, concrete blocks, or water can be used for weight on the platform.

Two test procedures are used in conducting the load test: the Slow Maintained Load Test Method, and the Constant Rate of Penetration Test Method.

12.6.1 Slow Maintained Load Method (SML)

The SML test is presently the most widely used procedure for determining the load-carrying capacity of a single pile. In the test, the load is applied in increments. Under each increment, the settlement of the head of the pile is observed as a function of time until the rate of settlement becomes very small. A new increment is then added. The results of typical SML load tests are shown in Figure 12.7, where the total load is plotted as a function of the settlement of the pile head. This procedure is recommended when estimates of settlements are required for working loads. The results are less reliable when ultimate loads are being determined.

Curve a represents a pile that plunged suddenly when the load reached a definite value termed "the ultimate pile load" or "pile capacity." Curves b and c show no well-defined breaks. The determination of the ultimate pile load is a matter of interpretation by the geotechnical engineer.

12.6.2 Constant Rate of Penetration Method (CRP)

The cost of a load test on piles depends greatly on the duration required to complete the test. This is especially important when the job is delayed while the results are pending. The duration of a load test can be shortened substantially by using the constant rate of penetration method.

In this method, the pile is forced into the ground using the hydraulic jack at a constant rate of 0.75 mm/min in clay soil and 1.5 mm/min in sands and gravels. The load is measured either by using a calibrated jack or by a load cell or proving ring. The load is increased until maintaining the specific rate of penetration requires no further increase in load. The corresponding total load is considered the ultimate bearing capacity.

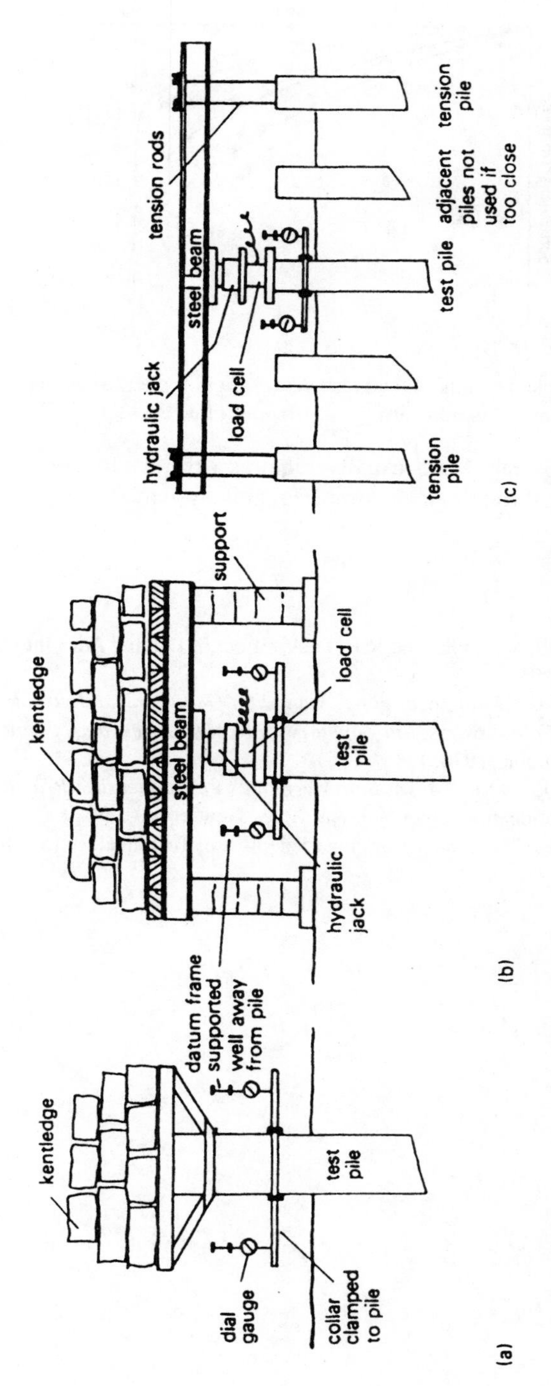

FIGURE 12.6 Pile loading test (a) kentledge only; (b) jacking against kentledge; (c) jack against tension piles (after Whitlow).

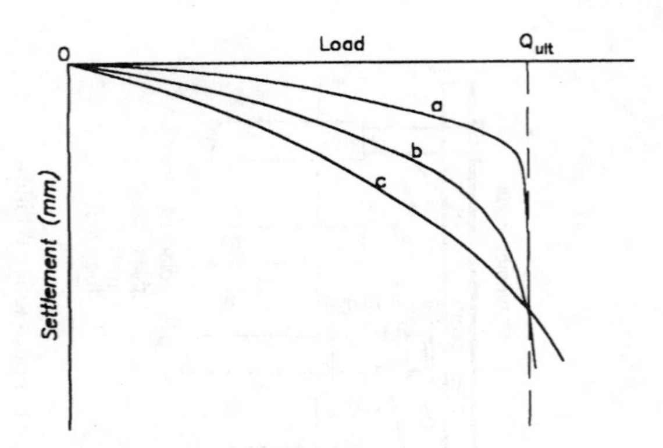

FIGURE 12.7 Typical results of load tests on (a) friction pile; (b) end bearing pile; (c) pile deriving support from both end bearing and friction (after Peck).

This method is rapid and usually requires several observers taking readings simultaneously, or through an electronic logging system.

REFERENCES

A.E. Cummings. G.O. Kerkhoff, and R.B. Peck, Effect of Driving Piles into Soft Clay, Trans. ASCE. 115, 1950.

R.B. Peck, Where has all judgment gone? *Canadian Geotechnical Journal,* 17, 1980.

G.B. Sowers and G.S. Sowers, *Introductory Soil Mechanics and Foundations,* Collier-Macmillan, London, 1970.

K. Terzaghi, R. Peck, and G. Mesri, *Soil Mechanics in Engineering Practices,* John Wiley-Interscience Publication, John Wiley & Sons, New York, 1996.

R. Whitlow, *Basic Soil Mechanics,* Longman Scientific & Technical, Burnt Mill Harrow, 1995.

13 Drainage

CONTENTS

At one time, Terzaghi stated "There will be no soil mechanics without water." With men on the moon, we understand that "lunar soil" is totally different from that on earth. Ironically, water is needed to stabilize soil, yet most construction failures are generated from too much water. Until recently, our knowledge of soil mechanics has been based on saturated soils. Indeed, only with rare exceptions can totally saturated soils be found; most soil on earth contains varying amounts of moisture.

Intensive studies on unsaturated soils were made by Professor Fredlund and others in Canada. Fredlund avoids the use of the term "partially saturated." He reasons that the soil is either saturated or unsaturated, analogous to the possibility of a lady being "partially pregnant."

Studies of drainage are generally limited to seepage and permeability. Details on drain installation, water disposal, bedrock geology, horizontal or vertical membranes, and others are usually looked at by academia as problems for construction engineers. The consultants fully realize that drainage problems on soil probably are more critical than some of the finite element studies that are taught in school.

In this chapter, drainage studies include the following:

1. The prevention of seepage out of the soil
2. The improvement of soil property
3. The decrease of swelling potential

4. The cut-off from moisture intrusion
5. The stabilization of slope
6. The subdrain around the basement

13.1 METHODS OF DRAINAGE

The movement of water through the soil mass is generally termed “seepage.” In engineering practice, soil is drained where it is desirable to eliminate seepage pressure or to increase the shearing strength. Drainage also is important when it is necessary to prevent seepage into the soil from exterior sources. Some of the methods of drainage are:

Drainage by gravity
Drainage by consolidation
Drainage by desiccation
Drainage by electro-osmosis
Drainage by suction

Of the above, the interests of the consultants are obviously in “drainage by gravity,” although soil suction on expansive soils has received increased attention.

13.1.1 Permeability

The method of drainage in an excavation of a construction project depends on the slope of the excavation, the permeability of the soils, the elevation of the water table, and the available equipment.

If the soil mass through which the seepage occurs is compacted fill, such as a dam or an embankment, the permeability can be reduced by the proper selection of material used. The classification of soils according to their coefficient of permeability is as shown in Table 13.1.

TABLE 13.1
Coefficient of Permeability of Various Types of Soils

Degree of permeability		Values of k (cm/s)
High	coarse gravel	Over 10^{-3}
Medium	sand, fine sand	10^{-3} to 10^{-5}
Low	silty sand, dry sand	10^{-5} to 10^{-7}
Very Low	silt	10^{-7} to 10^{-9}
Impervious	clay	less than 10^{-9}

In most soils, the coefficient of permeability depends on the direction in which water is traveling. In the direction parallel to the bedding planes, the value can be 2 to 20 times that in the direction perpendicular to the bedding stratification.

In the Rocky Mountain area, the claystone shale is generally impermeable. However, it contains numerous seams and fissures through which water can flow easily. The effective permeability will be far greater than that of the intact materials between the cracks. Laboratory permeability tests have little meaning with respect to the actual value. Consultants are warned to be very cautious in dealing with claystone since it may be the source of many foundation distress problems.

For major projects, it is advisable for the geotechnical engineers to seek advice from an experienced geologist and water engineers. Probably the most reliable method in determining the permeability of claystone bedrock is by conducting a field pump test.

13.1.2 SLOPE AND DRAIN

Slope modification and provision of drainage are positive methods for improving the stability of both cut and embankment slopes. Backfill around new construction is usually not well compacted and in many cases results in negative slopes. Surface water not only cannot drain away from the structure, but also accumulates around the foundation soil. Both the architects and the geotechnical engineers tend to ignore such conditions as their contracts do not involve such construction control.

For excavation drainage, methods that require pumping are commonly used. For permanent stabilization of slopes, a means of providing for gravity flow is necessary. Interception drains, ranging from farm-tile installations to elaborate forms of drain and filter constructions, are used.

Special attention should be directed to the installation of roof drains. It is a common practice to connect a roof drain outlet to the horizontal drain under the building. When the connection is clogged, all the water collected from the roof will drain into the foundation soils. Such conditions will not be detected for many years. As a result, the foundation can experience excessive settlement or unexpected heave.

The location of the drain system depends on the initial seepage pattern. If the drainage is for a building site, a highway, or a runway, the initial ground water condition must be established in order to determine the most suitable drain location; such an undertaking is sometimes ignored by the architect, and the entire design may be left to the contractor. Geotechnical engineers are often provided a stack of details by the architect, including the design of a door knob, but as far as the drainage detail is concerned, it is shown only by dotted lines.

A typical drain consists of three components: the filter, the collector, and the disposal system. It is a misconception that the pipe in the drain system is carrying water. Actually, it is the filter or the collector that is responsible for an efficient drain system. The filter is pervious enough to permit the flow of water into the drain with little head loss and at the same time fine enough to prevent erosion of the soil into the drain. A proper filter is the key to a successful drainage system. An improper filter, which is commonly used by contractors, is one factor that contributes to foundation failure.

For a filter to provide free drainage, it must be much more pervious than the soil. Figure 13.1 shows the grain-size criteria for material used as filter.

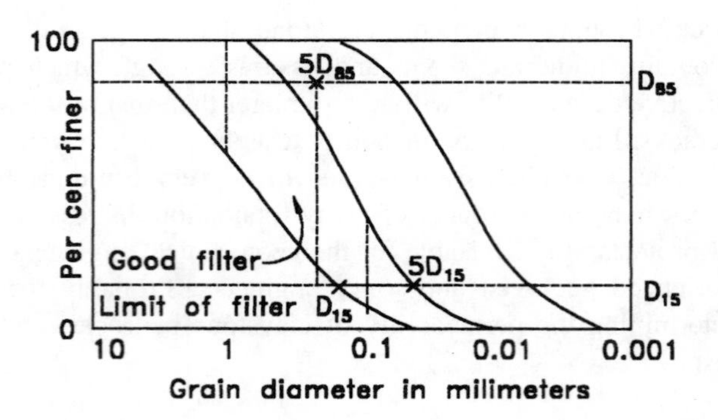

FIGURE 13.1 Grain size criteria for soils used as filter (after Sowers).

For many silty and clayey soils, a well-graded concrete sand makes a fine filter. In most construction projects, not many contractors will adhere to the above specification. The typical perforations in the drain pipe are 1/16 to 5/8 in.

More critical than the filter design is the outlet of the drain system. In the design stage, it is assumed that there will not be any obstacle in the outlet to prevent free flow. After years of service when the surrounding environment changes, and the outlet is blocked, such a condition will not be noticed until water from the drain backs into the building.

What contractors commonly refer to as a "French drain" is made of coarse gravel or crushed rock placed in a trench. Such a device is commonly employed by the contractor to take the place of a filter drain. A properly functioning "French drain" seldom exists.

13.1.3 Well Point and Pump Test

Well points are small-diameter wells that are driven or jetted into the ground. Usually they are placed in a straight line along the sides of the area to be drained and are connected at their upper ends to a horizontal suction pipe called a header, as shown in Figure 13.2.

Well points are usually spaced 6 to 30 ft. They are capable of lowering the ground water from 30 to as much as 100 ft. The cost of installation of such a system is high, and unless it is for a major project such as a large dam, the system is not warranted.

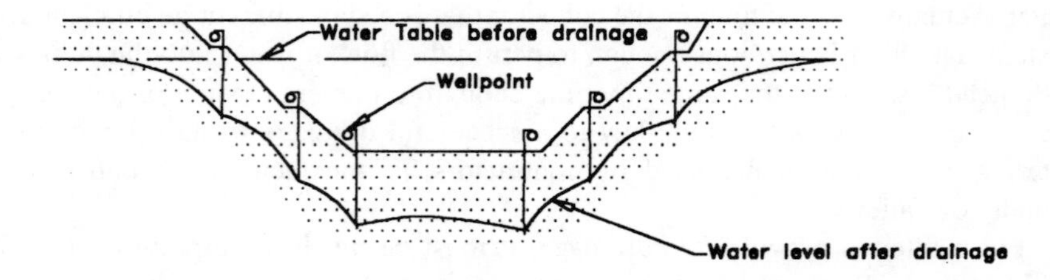

FIGURE 13.2 Multiple stage well point system (after Sowers).

Pump tests involve the measurement of a pumped quantity from a well together with observations of other wells of the resulting drawdown of the groundwater. A steady state is achieved when at a constant pumping rate the levels in the observation wells also remain constant. The pumping rate and the levels in two or more observation wells are then noted. The coefficient of permeability obtained from the field pump tests is far more reliable than the laboratory permeability tests.

The pump test is time-consuming, and its cost is high. By utilizing an on-site drilling rig and auger drills, the cost can be reduced. Pump tests should be performed on projects where water table and infiltration rates are critical.

13.2 GROUNDWATER

The term groundwater is loosely defined as a continuous body of underground water in the soil void that is free to move under the influence of gravity. The water table is the upper surface of a body of ground water. It is defined as the level of water in an open hole in the ground and is the level at which the pressure in the water is zero. Groundwater is not a static body but a stream with a sloping surface that takes many shapes, depending on the structure of the soils and rocks through which it flows.

13.2.1 Water Level

Geotechnical engineers commonly determine the water table level from the measurement of the water level in the exploratory boring holes. In sandy soils, such measurements are reliable. However, in clays soil or in claystone shale such readings can be misleading. Auger rotation tends to seal the surface of the clay preventing water from seeping through holes that field engineers usually log as "dry." In fact, as soon as overnight, water can appear in the boring holes. It is therefore important to check for water at least 24 h after drilling. Local ordinances sometimes require that the drill holes must be filled immediately after drilling. In such cases, the field engineer should fully qualify in the log the conditions under which the water level is measured.

The cost of drilling foundation pier holes depends a great deal on the groundwater level. The cost of drilling in wet holes, where dewatering is required, can be several times higher than drilling in the dry holes. When contractors bid the project, they base the descriptions on the field condition given in the field log. If the field log indicated "dry" holes and water is encountered, a "change of condition" will be submitted and the geotechnical engineer is blamed for the cost increase. It is amazing that the word "dry" can involve thousands of dollars.

Fluctuation of the groundwater level depends on the source of the supply. When the rate of intake exceeds the rate of loss, as it does during the wet season, the water table rises. When the intake decreases, as during the dry season, or when the loss increases because of pumping for water supply or because of drainage, the water table falls. Fluctuation of the groundwater level can reach as much as 10 ft.

If the time of the foundation investigation is during a low water table period and the penetration resistance of the soil is high, the foundation can suffer damage

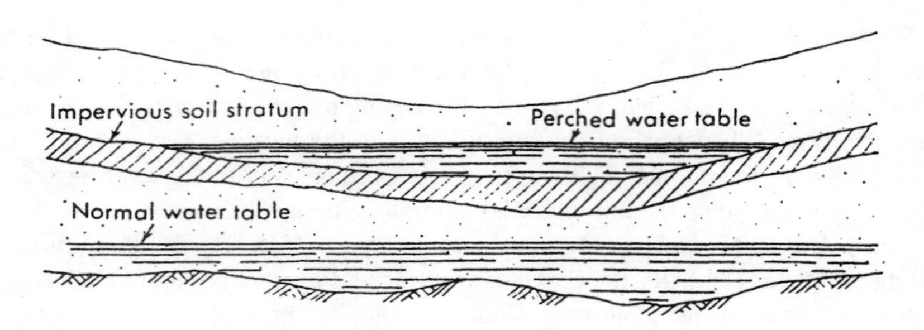

FIGURE 13.3 Perched aquifers (after Sowers).

when the water table rises and softens the soil. The problem may not emerge until many years after the structure is completed. A large percentage of foundation distress is caused by the rise of groundwater.

13.2.2 Perched Water

A perched water table occurs when water seeping downward is blocked by an impermeable layer of clay or silt and saturates the area above it, as shown in Figure 13.3.

An impervious stratum creates a basin that may hold groundwater that is perched above the general water table. A perched water table occurs rather frequently and is well recognized by geologists and water engineers. A small amount of perched water may be drained by drilling holes through the impervious basin, allowing the water to seep downward. However, in the Rocky Mountain region where claystone bedrock is near the ground surface, the extent of the perched water table can be very extensive. It poses a great many problems to the structure's foundation.

Perched water is fed by surface water derived from precipitation and snow melt. When the area is developed, perched water is further fed by lawn watering, drain from leaking sewer lines, and other man-made sources. A perched water reservoir can be replenished by a water source as far as a mile away. The size of a perched water reservoir can vary considerably. A small reservoir can pose a seepage problem only after a prolonged wet season, while some perched water reservoirs do not dry up even during dry seasons.

With the development of a perched water condition, water from the reservoir can gradually seep into the foundation soil and cause extensive damage, especially in the expansive soil area.

13.2.3 Moisture Barrier

Moisture barriers are used to prevent groundwater or water from other sources from seeping into the excavation. Both vertical and horizontal barriers have been used. Theoretically, vertical barriers should be more effective than horizontal barriers in minimizing seasonal drying and shrinking of the perimeter foundation soils, as well as maintaining long-term uniform moisture conditions beneath the covered area.

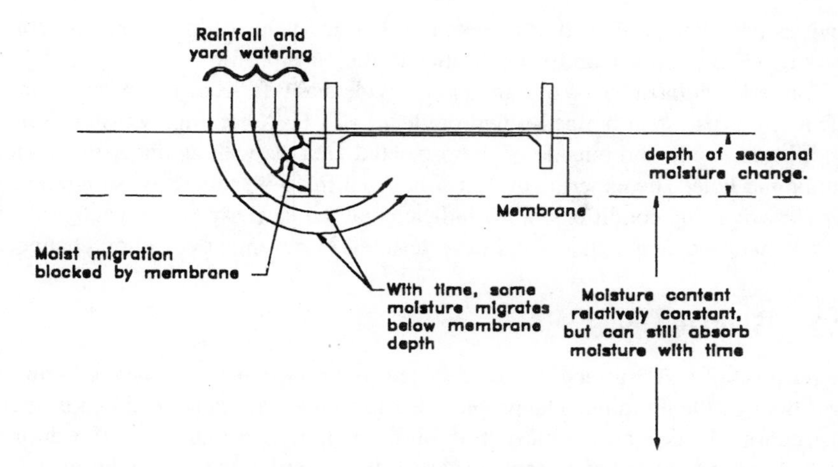

FIGURE 13.4 Path of moisture migration blocked by vertical barrier.

Indeed, horizontal membranes can be placed around the building and under the lower slab, but the long-term effect of intercepting water is limited. At the same time, the result from the use of deep vertical barriers in intercepting water is encouraging. Buried vertical barriers may consist of polyethylene membrane, concrete, or other durable, impervious materials. The path of moisture migration when using vertical barriers is shown in Figure 13.4.

Vertical moisture barriers should be installed to a depth equal to or greater than the depth of seasonal moisture changes. The following should be considered in the installation of a moisture barrier:

1. Vertical moisture barriers are difficult to install around basement structures, due to the deep ditch excavation required.
2. Vertical moisture barriers should be installed 2 to 3 ft from the perimeter foundation to permit machine excavation of the trench for the membrane.
3. A vertical moisture barrier is sometimes attached to a horizontal moisture barrier to prevent wetting between the vertical barrier and the building.
4. Either concrete or polyethylene membrane can be used. The membrane should be of sufficient thickness and durability to resist puncture during backfilling of the trench.
5. It is also possible to use semi-hardening impervious slurries installed in a narrow trench instead of membranes.

Probably the most successful and well-documented use of a vertical moisture barrier to control pavement distress is the experiments in San Antonio, Texas, by Malcolm Steinberg. Since horizontal moisture barriers over subgrade cannot be applied to existing pavements, the use of a vertical moisture barrier is a logical solution.

The soil under the test section consists of high plasticity clays with a liquid limit from 50 to 72. The highway was built in 1970. Since its construction, subgrade activities have been reflected in repeated asphalt level-ups of the pavement, continuous surface distortions, and irregularities in the curb profiles.

The test sections consist of a trench 3 ft wide and 9 ft deep. The barrier consists of DuPont Typar span bonded polypropylene with EVA coating. Moisture sensors were placed inside and outside of the protected area, as well as under the adjacent southbound lane. The experiment was conducted in 1979; today, the subgrade soils have shown drying conditions. They indicate that the barrier-protected subgrade had fewer moisture changes that could have resulted in pavement elevation changes.

13.3 SUBDRAIN

The purpose of a subsurface drainage system is to prevent water from entering the lower floor of the building. The problem is especially serious for residual basement construction. In order to achieve this purpose, it is necessary that the drain be installed with adequate depth, capacity, and a proper outlet. In the past, little attention has been given to the detailed design of the subdrain system. A great majority of the drain systems will not function. In a single-family home, occupants often find that water gradually seeps into their basement several years after the completion of the house.

In terms of cost, the problem is also significant. In 1978, it was estimated that the cost to install a subdrain system for a single-family house in metropolitan Denver was about $250. For 20,000 homes built in the entire state, the cost of subdrain system amounted to $5 to 7 million. Today, the cost can easily be doubled. Certainly, with this kind of money, the design and construction of the system merits a critical review.

13.3.1 Installations of Subdrain

The most significant source of water that enters into a residential basement is from lawn watering. Home owners tend to over-irrigate their lawns. A horticulturist claims that the water required to maintain a healthy lawn is only less than half what is generally used. The excess water will eventually find its way into the foundation soils.

Another common practice is to place a plastic membrane over the backfill area and cover with decorative rocks. The purpose of such an installation is to prevent water from entering the backfill and to prevent the growth of weeds. Such practices serve their purpose for a few years, then water enters beneath the plastic through the joints. The water beneath the plastic does not evaporate and the amount increases gradually.

Another main source of water that can enter into the basement is the perched water. A perched water condition will not take place until the area is fully developed and most buyers will not suspect the development of a perched water condition until it is too late.

Usually, there is no significant advantage in placing the drain outside the building. With the drain placed inside the perimeter of the foundation walls, the masons are able to place the drain more accurately with respect to the floor level. As a result, a better geometry of the system can be achieved. In addition, the drain can be exposed by cutting the concrete slab for inspection at a later date if required. Also, with the drain placed inside the basement, gravel in the trench can be easily connected with the underslab gravel. A combination of subdrain and underslab gravel should constitute a most effective drain system.

It is important that the drain tile be located a considerable distance below the floor level. The deeper the location of the drain, the further the drawdown effect will be and the better will be the protection offered to the basement. Unfortunately, cost will not allow the construction of a deep drain system. In most cases, the typical peripheral drain will not be able to draw down a perched water table to prevent a wet basement condition.

Experience indicates that the drain tile should be installed at least 12 in. below the top of the slab. A depth of 24 in. is preferable for large basements or where the perched water table is near floor level.

13.3.2 Drain Outlet

The gravity outlet provides the most inexpensive system, with practically no maintenance problem. With favorable topographic features, there is no doubt that the gravity outlet offers the best solution. However, the gravity outlet may be plugged as shown in Figure 13.7.

For residential basement construction, it is common practice to discharge water from the subdrain into the storm sewer. Since it is not legal to discharge water into the sanitary sewer, the alternative scheme is to provide gravel beneath the sanitary sewer and connect the subdrain into the under-pipe gravel.

First, it is possible that there is an insufficient amount of gravel beneath the storm sewer line to carry away the water. In some extreme cases, gravel is absent for several feet, and thus, the defective system will not carry any water. Second, the drop between the under-pipe gravel and the under-slab drain line may not be sufficient to allow gravity flow.

With the entire system covered, the owner has no way to know about the faulty installations until some major problem emerges.

In addition, such a system usually cannot effectively protect the basement from the rise of groundwater or a perched water condition. The drain is located in close vicinity of the floor slab that does not allow a significant drawdown effect. Considering that most wet basement conditions are due to the development of a perched water condition, it is doubtful that the near-surface drain system can solve the problem.

Directing flow from the subdrain system into a sump and then pumping it out offers a most desirable system where a positive gravity outlet cannot be achieved. The drain outlet should extend to about 3 ft below the sump outlet. Pumped water can be discharged to the storm sewer or sometimes the gutters along the street. Theoretically, sump pump installations are suitable only for areas where the coefficient of permeability is low and where only manageable inflow of water enters the system. Figure 13.5, prepared by the Federal Insurance Administration (FIA), indicates that the sump pump system is adequate when the inflow is less than 45 gpm,

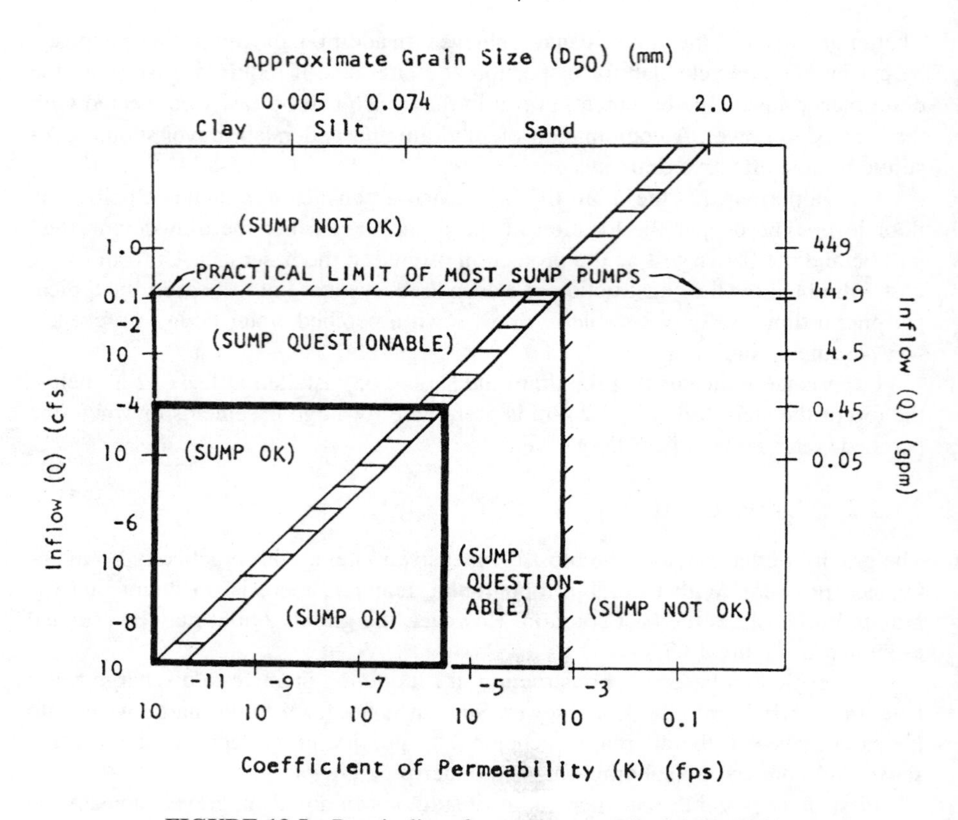

FIGURE 13.5 Practicality of sump pump method (after FIA).

and the coefficient of permeability is less than 1/1000 ft/sec. or 32,000 ft/yr. In most cases, a 1/3 to 1/2 h.p. pump should be adequate for residential use.

The least desirable type of water disposal method is by the installation of a "dry well." In general, dry wells can be used to dispose of water only in granular soil areas where the rate of percolation is high. For a cohesive soil with low permeability, the capacity of a dry well is limited. When the well is flooded, not only will its function be terminated, but water will also be backed into the drain system, causing severe damage to the structure.

13.3.3 Rational Design

This chapter has little to do with the mechanics of soil. Most textbooks on soil mechanics deal only briefly with this subject as they involve no mathematical treatments. In reality, in most cases of foundation damage, a water or drainage problem is directly or indirectly involved (see Figures 13.6 to 13.9).

For a rational drainage design, consideration should be given not only to its effectiveness, the construction capability, and the existing regulations, but also to the cost involved. The following points should be considered:

FIGURE 13.6 Improperly placed drain outlet.

FIGURE 13.7 Partially blocked drain outlet.

FIGURE 13.8 Cobble above plastic membrane over backfill plank preventing drainage.

FIGURE 13.9 Down spout drain into backfill.

1. Prior to construction, the possible source of water that may affect the stability of the structure must be studied. The water level taken from the drill holes can only be used as a reference.
2. The possibility of the development of a perched water condition at a later date must be considered.
3. A subdrain system can be installed either inside or outside the structure. For residential construction, it is desirable to place the system around the inside perimeter of the basement, so that the system can be connected to the underslab gravel.
4. All subdrains should be placed at least 3 ft below the lower floor level.
5. The practice of connecting the subdrain into the gravel under the existing sanitary sewer line should be abandoned.
6. The subdrain outlet must be inspected and maintained to ensure a free flow of water.
7. The total length of the drain line should not exceed 1000 ft, with adequate slope before an outlet is provided.
8. There should be sufficient gravel thickness under the drain pipe to allow for the volume of water.
9. It is advisable to provide a sump pit in the lower floor of all construction so that in case of emergency, a pump can be inserted.
10. The cost of providing a gravel layer under the lower floor slab is justified.
11. The practice of using a "dry well" should be discontinued.

13.4 CASE EXAMPLE

In 1984, a developer proposed building a subdivision in southwest Denver. The location was known to builders and engineers as a potentially high-risk area. The soils possess high swelling potential. The developer obtained a comprehensive soil report from a qualified geotechnical engineer. In the soil report, the following facts were pointed out:

The subsoil consists of shallow overburden soils over claystone bedrock.
The site is underlain by steeply dipping cretaceous sedimentary rocks.
No free water was found in the test holes to a depth of 35 ft, 74 days after drilling.
There is a risk of developing a future perched water condition on top of bedrock due to irrigation within the development.
There is a five-acre pond located about a mile to the south of the development, as shown in Figure 13.10.

In 1989, when the development was partially completed, many of the houses suffered damage and water entered into the basements. Extensive investigation was conducted to determine the source of water that entered into the area. After months of intensive study by college professors, geotechnical engineers, geologists, and water engineers, two schools of opinions were established: that water was derived

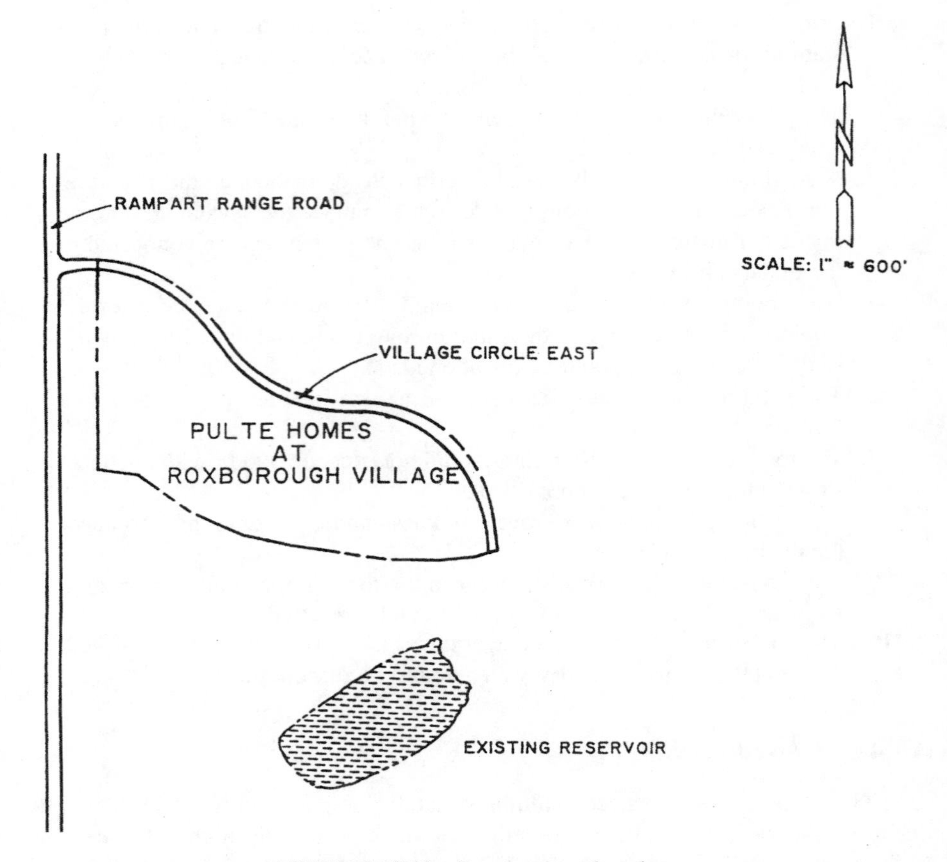

FIGURE 13.10 Development location.

from the perched water condition, and that water was derived due to seepage from the pond.

13.4.1 Perched Water

This opinion claimed the water that caused damage to the residences was primarily due to the development of a perched water condition. The following are the main arguments:

1. Prior to the development, no free water was found to a depth of 35 ft. After the development, water was found at a shallow depth in the area of the completed houses.
2. As shown in Figure 13.11, at the vacant lots to the east of the development the soils maintained the initial moisture content with no change of water level.
3. A perched water condition was created by the lawn watering practice used by the owners of the completed houses. It is also possible that faulty

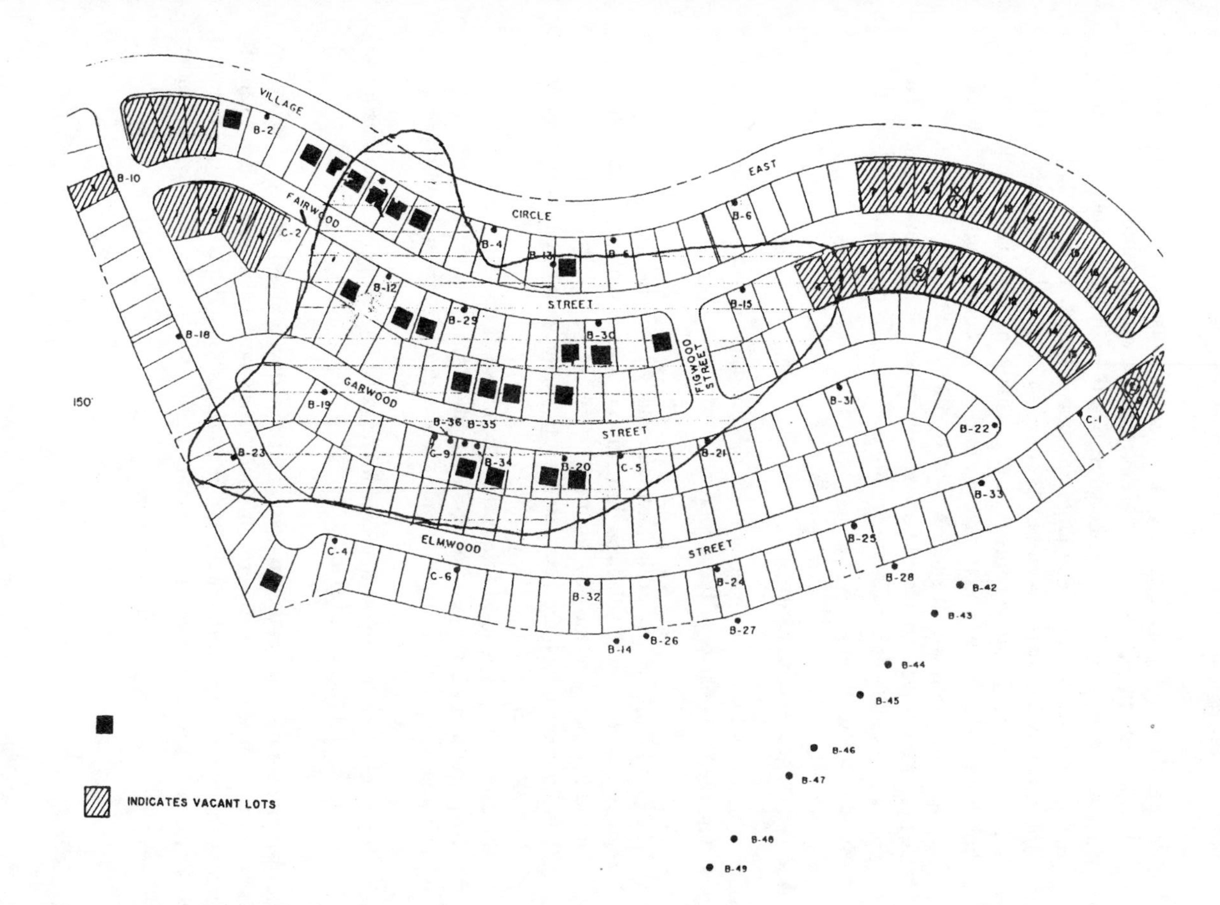

FIGURE 13.11 Partially completed development (perched water and damaged houses).

constructions of the subdrain system contributed to the volume of water in the perched water reservoir.

4. Figure 13.11 approximately delineated the perched water boundary. It is hardly coincidental that most of the damaged houses are located within this area.
5. Prior to the development, there was a natural balance between evaporation and transportation. After an area is developed, the environment is changed. Covered areas, such as walks and streets, roof drainage, and others destroy the natural balance and feed large amounts of water into the soil.
6. The excess water can seep into the upper soil and will not penetrate into the lower hard claystone bedrock, and gradually builds up a perched water condition. The process may take several years before turning the once-dry area into one with a shallow water table.
7. A deep trench was excavated between the pond and the development. No free water was found, indicating that the seepage from the pond must only be in a small amount.

13.4.2 Seepage from Pond

Another opinion is that water seeped from the pond has caused the damage to the houses. The theory consists of these factors:

1. The pond is not lined and leakage can easily take place.
2. Seepage water can travel through the upper soils and through the fissures in the bedrock into the development.
3. It is estimated that the seepage loss from the pond is between 10 to 80 gallons per min.
4. Suction theory was used to advance the cause of damage.
5. Diffusion theory was used to predict foundation movement.
6. Infrared mapping was used to trace the seepage path from the pond to the development.
7. It was claimed that the water level in the pond had dropped. The question is, where did the water go?

After months of study and deliberation the key question — "Would there be any damage to the houses if there were no pond?" was never answered. That example indicates the importance of drainage in soil engineering and the difference between theory and practice. It also verifies the concept that soil mechanics is an art rather than a science.

REFERENCES

F.H. Chen, *Foundations on Expansive Soils,* Elsevier Science, New York, 1988.

Manual for the Construction of Residential Basements in Non-Coastal Environs, Department of Housing and Urban Development, Federal Insurance Administration, Washington, D.C.

G.B. Sowers and G.S. Sowers, *Introductory Soil Mechanics and Foundations,* Collier-Macmillan Limited, London, 1970.

M.L. Steinberg, *Geomembrance and the Control of Expansive Soils in Construction,* McGraw-Hill, New York, 1998.

D.K. Todd, *Ground Water Hydrology,* John Wiley & Sons, New York, 1980.

14 Slope Stability

CONTENTS

Slides can take place either in a natural formation or in a man-made mass. Theoretically, sliding can occur if the shear stresses developed in the soil exceed the corresponding shear strength of the soil. Stability analysis is important in many kinds of projects, such as highway cuts and fill, earth dams, canals, basements, etc.

The movement of a large mass of soil usually takes place along a definite surface and can be easily identified. As the slide continues, it becomes distorted and broken up and may be difficult to recognize. Movement can take place suddenly, while some takes place after showing such signs as tension cracks. It is difficult to determine the cause of slope failure, although generally it is associated with the increase of moisture content. In the early days, slope failure was always written off as "an act of God." Today, attorneys can always find someone to blame and someone to pay for the damage — especially when the damage involves loss of life or property.

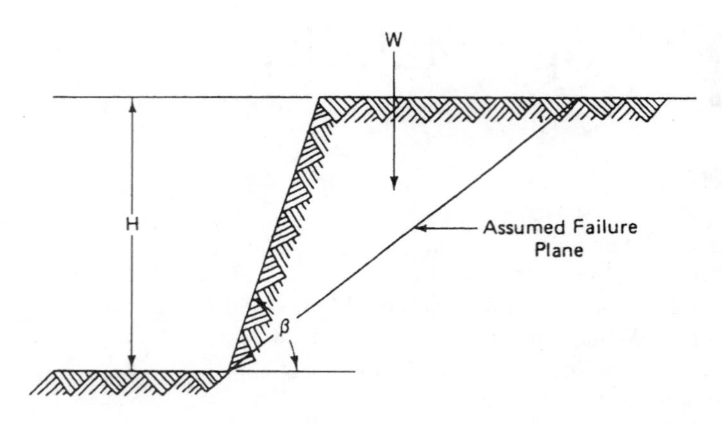

FIGURE 14.1 Failure Plane (after Liu).

14.1 ANALYSIS METHODS

Various methods of stability analysis are used by engineers. Some use theorems based on the shape of the slide, others base their analysis on soil properties. The various methods are briefly summarized below:

14.1.1 Culmann Method

As early as 1875, long before soil engineering was known, C. Culmann of Switzerland devised a graphical method to analyze the stability of a slope consisting of homogenous soils. Culmann's method originated from the classical earth pressure theorem of Coulomb. His method is based on the assumption that the failure of a slope occurs along a plane when the average shearing stress tending to cause the slip is more than the shear strength of the soil. It is assumed that failure occurs along a plane that passes through the toe of the slope, as shown in Figure 14.1.

The safe depth of cut, H, can be computed for a given soil as a function of the angle β.

14.1.2 Taylor's Stability Number Method

The stability number N is defined as

$$N = \gamma H/c$$

where γ = unit weight
H = height of cut
c = cohesion

As shown in Figure 14.2, three types of failure surface are possible: base failure, toe failure, and face failure.

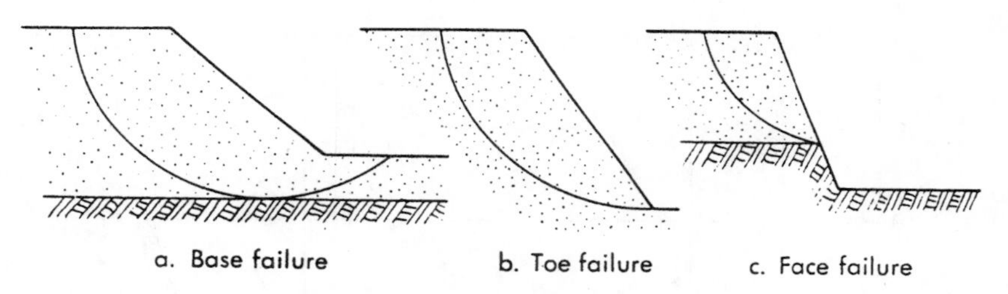

FIGURE 14.2 Types of slope failures (after Sowers).

In the case of the base failure, the surface of the sliding passes at some distance below the toe of the slope. Base failure generally develops in soft clays and soils with numerous soft seams. The top of the slope drops, leaving a vertical scarp, while the level ground beyond the toe of the slope bulges upward. Toe failures occur in steeper slopes. The top of the slope drops, while the soil near the bottom of the slope bulges outward. Face failures occur when a hard stratum limits the extent of the failure surface. The type of slope failure engineers most frequently encounter is the toe failure.

In many regions, railways or highways must be built on swampy ground or in broad valleys filled with soft clay or silt, where the fill often sinks bodily into the supporting soil. Failure of an embankment due to excessive settlement may take place. Such an occurrence is referred to as "failure by sinking." On the other hand, the fill together with the layer of soil on which it rests may spread onto an underlying stratum of very soft clay or water under pressure. Such a case is known as "failure by spreading." The new Hong Kong airport with runways extended well into the sea must have taken such cases into consideration.

In the case of "toe failure," the failure surface passes through the toe. The stability number can be determined for a specific case by the value of the angle of internal friction and the slope angle. If the angle of internal friction is zero or nearly zero, such as plastic clay or silt, Figure 14.3 can be used to determine both the types of failure surface and the stability number. From the figure, it can be seen that toe failures occur for all slopes steeper than 53°. For slope angles less than 53°, there is the possibility of slope failure, depending on the value of stability number. When the stability number is 3 or more, a base failure will occur. For stability numbers between 1 and 3, a base failure or a toe failure may take place, depending on the slope. For a stability number less than 1, only slope failure can take place.

14.1.3 Method of Slices

In order to compute the stability slopes in soils whose strength depends on the normal stress, it is necessary to compute the effective normal stress along the failure surface. Fellenius in 1920 developed an approximate solution. The method of slices is to draw to scale a cross-section of the slopes as shown in Figure 14.4.

A trial curved surface along which a sliding failure is assumed to take place is divided into a number of vertical slices. The weight of soil within each slice can be calculated by multiplying the volume of the slice by the unit weight of the soil. This

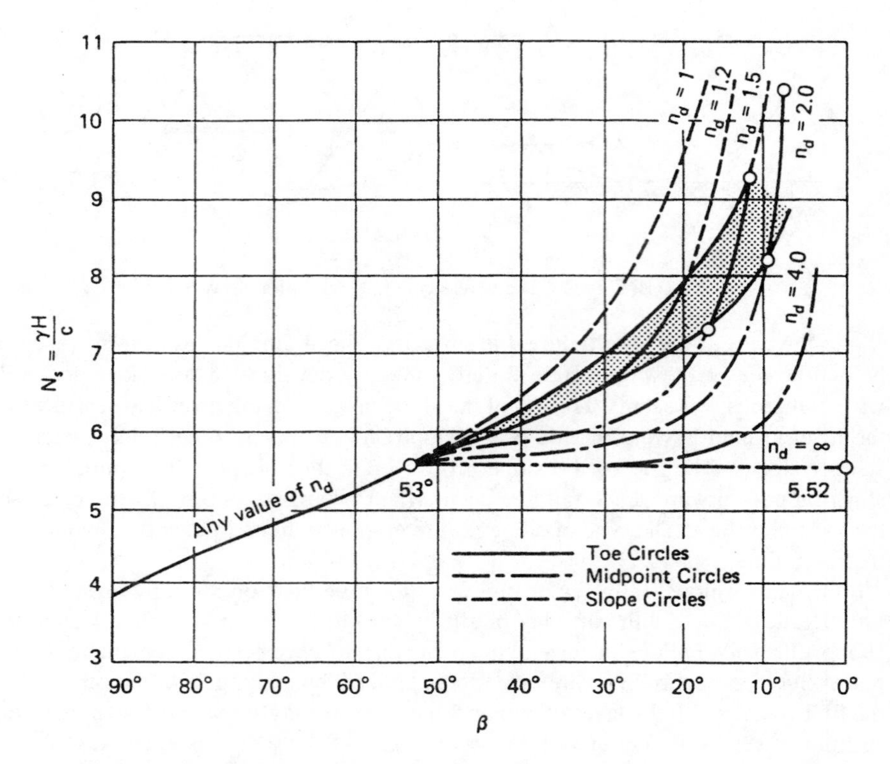

FIGURE 14.3 Stability numbers and types of slope failures for $\phi = 0$ (after Taylor).

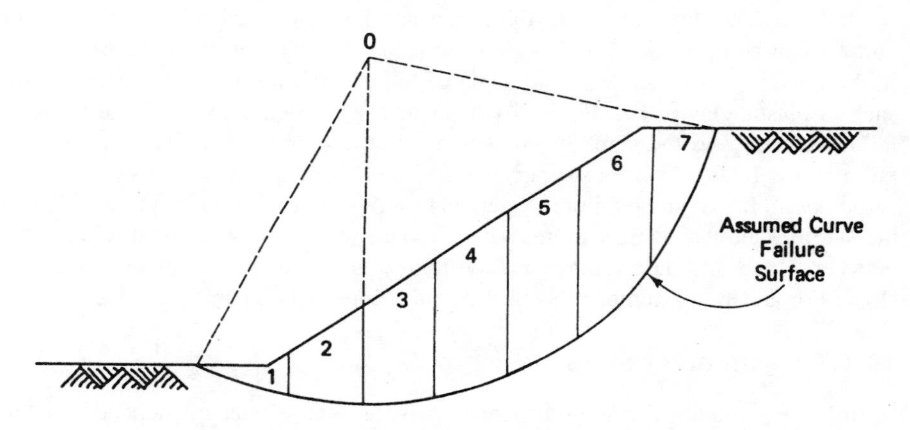

FIGURE 14.4 Circular segment divided into slices (after Liu).

force can be resolved into a component normal to the base of the slice and a force parallel to the base of the slice, as shown in Figure 14.5. It is the parallel component that tends to cause the sliding failure. Resistance to sliding is provided by the cohesion and the friction of the soil. The cohesion force is equal to the cohesion of

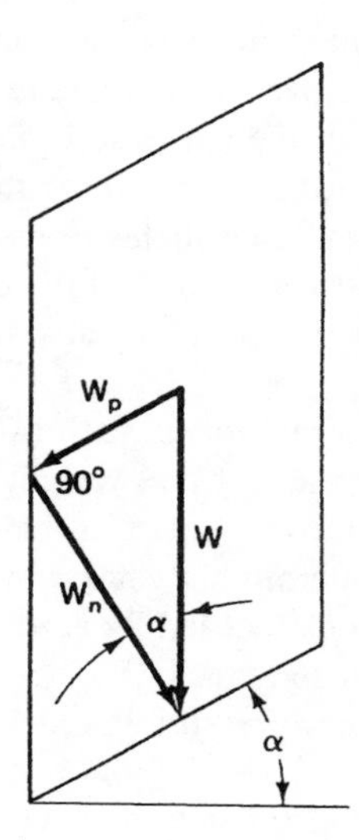

FIGURE 14.5 Forces acting on slice (after Liu).

the soil multiplied by the length of the curved base of the slice. The friction force is equal to the component of W normal to the base (Wn) multiplied by the friction coefficient (tan ϕ).

The total force tending to cause sliding of the entire soil mass is the summation of the products of the weight of each slice times the respective value of sin α, or ΣW sin α. Since Wn is equal to W multiplied by cos α, the total friction force tending to resist sliding of the entire soil mass is the summation of the products of the weight of each slice times the respective value of cos α times tan ϕ, or ΣW cos α tan ϕ. The total cohesion force tending to resist sliding of the entire soil mass can be computed by multiplying the cohesion of the soil by the total length of the trial curved surface, or cL. Based on the foregoing, the factor of safety can be computed by the equation

$$\text{F.S.} = \frac{cL + \Sigma\ W \cos\alpha \tan\phi}{\Sigma\ W \sin\alpha}$$

The analysis requires trial computation to a large number of assumed failure surfaces. The failure surface that has the smallest factor of safety is the most critical surface, or where the slide is most likely to take place. With the application of the computer, the process can be easily accomplished.

Many refinements and a variation of stability analysis have been proposed. This includes the circular arc analysis and the widely used Bishop's simplified method and others. While none of the methods are rigorous, they are sufficiently accurate for analysis and design.

14.1.4 Limitation of Stability Analysis

All methods of stability analysis are based on the assumption that within the failure surface the soil is homogenous. In practice, such conditions seldom exist. Furthermore, since movements occur when the shear strength of the soil is exceeded by the shear stresses over a continuous surface, failure at a single point cannot represent the failure of the entire soil mass. It is hard to determine the cause of the earth movements. Actually, anything that results in a decrease in soil strength or an increase in soil stress contributes to instability and should be considered in both the designs of earth structures and in the correction of slope failures.

Sower listed the causes of increased stresses as follows:

External loads, such as buildings, water, or snow
Increase of unit weight by increased water content
Removal of part of the mass by excavation
Undermining caused by collapse of underground caverns or seepage erosion
Shock, caused by earthquake or blasting
Tension cracks
Water pressure in cracks

He also listed the causes of decreased strength as follows:

Swelling of clays by adsorption of water
Pore water pressure
Breakdown of loose or honeycombed soil structure with shock, vibration, or seismic activity
Hair cracking from alternate swelling and shrinking or from tension
Strain and progressive failure in sensitive soils
Thawing of frozen soil or frost tension
Deterioration of cementing material
Loss of capillary tension on drying

Failure can result from any one or any combination of the above factors. Geotechnical engineers are often called to determine the cause of failure. It is easy to name the probable cause or causes, but it is difficult to prove the reasoning.

14.2 STABILITY OF LOESS

Loess is a semi-plastic wind-laid silt consisting chiefly of angular and sub-angular grains with an effective grain size between about 0.02 and 0.006 mm and a low uniformity coefficient. Loess is relatively stiff at its natural moisture content but may undergo a radical decrease in volume upon wetting. Geotechnical engineers classify loess as a collapsible soil.

Thick beds of wind-blown silt accumulate over the years on semi-arid grasslands. The deposits build up slowly, probably over centuries; therefore, the grass growth kept pace with the deposition. The result is high vertical porosity and vertical cleavage with a loose structure.

Most loess soils are hard because of the deposits of calcium carbonate and iron oxide that line the former root holes, but they become soft and mushy when saturated.

The cohesion is due to thin films of clay or cementing material that cover the quartz grains and the walls of the root holes. Since the root holes are predominately vertical, loess tends to break by splitting along the vertical surface.

Another explanation is that due to the high vertical permeability of loess, water tends to seep into the soil in the vertical direction. The breakdown of the cementing film in loess allows further seepage of water into the material until a vertical surface is created. The mechanics of the formation of loess deposit are not fully understood.

14.2.1 Cut Slopes

When loess is located permanently above the water table, it is a very stable material. To reduce the effect of erosion due to rain and snowmelt, the slope of a cut in loess should maintain a near-vertical position.

Deposits are found in China, Turkestan, and part of Nebraska. The origin of loess in different countries may not be the same. It is well known that if loess is subjected to prolonged wetting, the material will collapse, resulting in a very soft mass. An experiment in Turkestan where the embankment of a canal consisted of loess showed that it did not take long before the embankment collapsed.

At the same time, engineers in China performed railroad and highway cuts with vertical slopes in loess with great success. It is common knowledge that the slope of cut in loess should be vertical, so that precipitation will not accumulate on the slope.

14.2.2 Loess Deposit in China

In China, loess covers an area of over 200,000 square kilometers and is an important natural feature in the landscape of Western China. The depth of the deposit can reach as much as 200 ft, as shown in Figure 14.6.

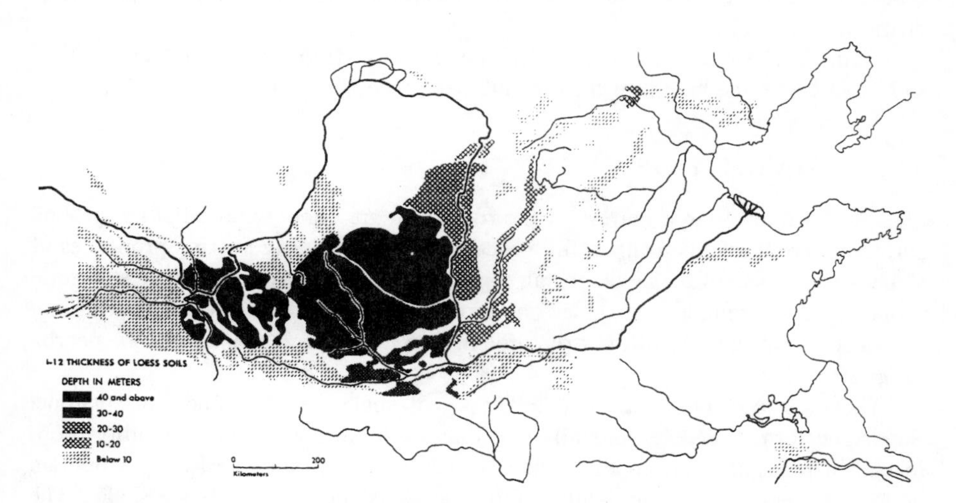

FIGURE 14.6 Distribution and thickness of loess deposit in China (after Hsieh).

FIGURE 14.7A Typical loess deposit embankment.

The accumulation of loess began when North China was mostly grassland. Apparently, strong outblowing winds from the Gobi Desert deposited silt on the grasslands that eventually were buried under these aeolian deposits. Today, the "sandy winds" or dust storms in spring in North China indicate that loess accumulation is still in progress.

In addition to cultivation, the most notable human use of loess is for cave dwellings that are excavated from the loess banks. These caves are cool in the summer and warm in winter, providing a reasonably comfortable home. Even couches and tables can be made of compacted loess.

In Northwestern China, highway alignment is such that it is necessary to cross many deep gullies. These gullies are as deep as 100 ft or more and are usually dry, with the exception of occasional rain storms. Instead of building costly bridges, highway engineers fill the gully with compacted loess, then tunnel through the bottom of the fill to allow water passage. Such temporary measures serve the flow of traffic very well.

Figure 14.7A is a typical embankment in the loess deposit area with near-vertical slopes and with the height of the embankment exceeding 50 ft.

14.3 STABILITY OF SHALE

Shale constitutes a large portion of the rocks that are either exposed at the ground surface or are buried beneath a thin veneer of sediments. Engineering properties of shale are of concern to geotechnical engineers since a large number of structural foundations are founded on shale. The movement of shale mass upon wetting not only affects the stability of the structure but also the life and property of the nearby residents.

The most commonly encountered types of shale are claystone and siltstone. Sandstone formations are generally stable and seldom pose slope instability problems. Siltstones and claystones are hardened silts and clays. In the Rocky Mountain area, such deposits are commonly referred to as "shale" or "claystone shale." The unweathered, deep-deposited claystone is used to support drilled piers. Foundations

FIGURE 14.7B Cave dwelling in loess deposit.

for high-rise buildings are placed on shale with piers designed with a high-bearing capacity. Claystone or siltstone, when exposed at the cut slope, can sometimes cause many stability problems to the highway engineers.

14.3.1 Near Surface Shale

Shales are commonly weakened by a network of joints. At lower depths, the joints are closed and are widely spaced. However, as the depths of overburden are decreased, the joints open because of the unequal expansion of the blocks located between them. The water content of the blocks then increases and their strength decreases. Slope failures either on hill sides or in cuts occur only during the rainy season. If a failure occurs, the slide material sometimes flows for a short time as a viscous liquid and then comes to rest.

Such slides can sometimes be stabilized by the installation of a horizontal drainage ditch along the head of the slide. Care should be exercised to ensure that the ditch should be deep enough to pass the unstable shale deposit. Otherwise, sometimes the ditch will encourage the slide. The resulting slide is illustrated in Figure 14.8.

The Bearpaw formation, consisting of layers of bentonite formed by the weathering of volcanic ash, interspersed with the more usual shales, is found in the western U.S. At the site of large open excavations, the rebound of the ground due to unloading and the swells of the bentonite layer tend to weaken the mass and trigger the slide. At some dam sites in Wyoming, extremely flat slopes had to be adopted.

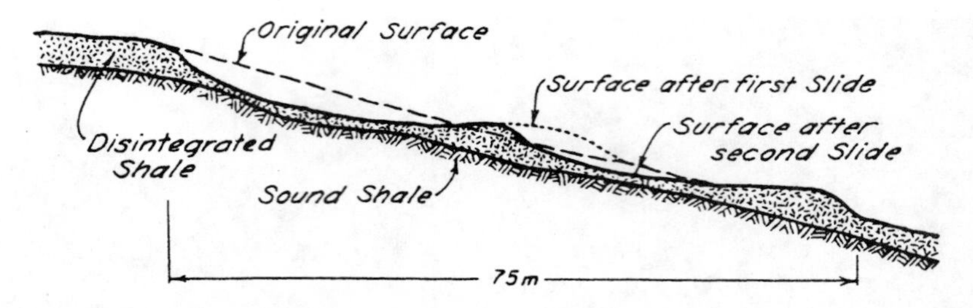

FIGURE 14.8 Profile of double slide in well bonded shale (after Ladd).

With only a small increase in water content some of the silty shales will slide. Slope failures, either on cuts or on hillsides, occur only during rainy season. The slide is relatively shallow, no more than 15 ft. Typical slide topography is shown in Figure 14.8.

14.4 NATURAL EARTH MOVEMENT

Slides in natural soil mass may be caused by man-made disturbances such as undercutting the foot of an existing slope by digging an excavation with unsupported sides. On the other hand, natural soils may move without external provocation on slopes that have been stable for many years. Most natural soil movement takes place after heavy rain or flooding. Prolonged drying and wetting, and a loss of vegetation cover or earthquake can trigger slides.

Potential earth movement areas are difficult to determine. The development of tension cracks at the head of a slope generally gives good warning. Slides that took place many years ago sometimes can be detected by observing the distortion of the tree trunks.

There are four different classes of natural soil movements:

1. **Creep** — The slow, relatively steady movement of soil down slopes
2. **Landslide** — The fairly rapid movement of soil or rock masses in combined horizontal and vertical directions
3. **Rockfalls** — The vertical superficial rock movements
4. **Subsidence** — The movement of the soil masses vertically downward

14.4.1 Creep Failure

Creep is a slow, nearly continuous movement of soil resembling the creep of metals under small stresses or the plastic flow of concrete. Creep failure was recognized by Terzaghi as early as 1935 when he stated "that the strength of soil can decrease with time and progressive failures occur." Since then, the creep phenomenon has been researched by many scholars of soil mechanics. It is interesting that when creep has been considered, it usually has not been applied to the ultimate failure of the slopes, and when slope failures have been considered, creep behavior usually has been ignored.

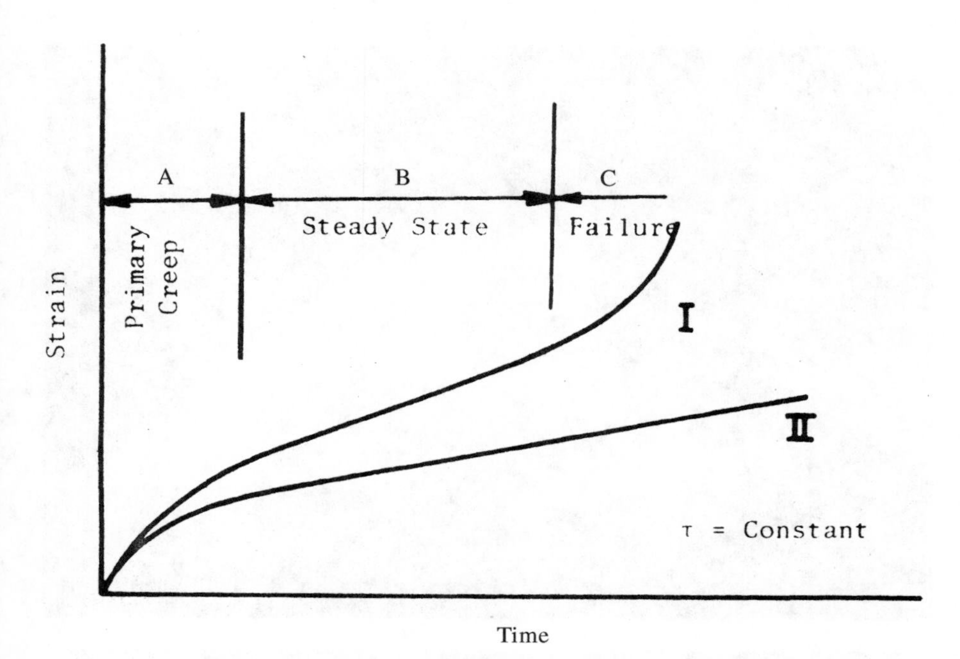

FIGURE 14.9 Typical creep curves (after Mitchell).

Casagrande, during his investigation of the Panama Canal failure, noted that the long-term strength of the soil was considerably less than the short-term. It was noted that the strength mobilized from the effect of creep is between the peak value and the residual strength of the clay. Skempton showed that progressive failures continue until a stable condition exists, at which point the stress of the creep is less than the residual strength.

Mitchell showed that for sufficiently high stress levels, creep rupture can occur, but for low stresses the creep reaches a constant rate with no rupture. Figure 14.9 indicates typical creep curves.

Creep movement is somewhat analogous to that of bricks placed on a tin roof with a slight pitch. Over a period of weeks or months, the bricks may be observed to be moving slightly down the roof. Considerably more research is needed to understand the mechanics of creep. To a consultant, damage caused by creep is less critical than that of landslides, yet its existence cannot be ignored.

14.4.2 LANDSLIDE

A number of slides in nature soils are of interest to the engineers, but which they find difficult if not impossible to analyze quantitatively. Natural slopes on steep hillsides that appear to be on the verge of sliding down may stand for a long time. On the other hand, stiff clay or shale on slightly sloping ground may present a great deal of trouble to the engineers.

The most common cause of landslide is seepage. Almost all major landslides in nature soils are associated with drainage. The rise of ground water, change of

FIGURE 14.10 Typical landslide along a highway cut area.

drainage patterns, deforestation, flood, excessive rainfall, or earthquake-induced flooding can cause massive or localized landslides. Some of the smaller-magnitude landslides can be stabilized by improving drainage. Whenever possible, engineers should try to avoid the potential slide risk area. A typical major landslide along a highway in California is shown in Figure 14.10.

During the location of the Burma Road, due to insufficient time for detailed survey, a section of road was located in a potential landslide area. After the road was open to traffic, a massive slope failure took place. At one time, as many as six slides in the cut slope occurred on the same day. Clearing the slides became a major task. At last, the engineers gave up and – at great expense – relocated 10 miles of the existing road to a stable area.

A granular soil that is looser than the critical density may pass into a state of complete liquefaction if failure starts. Some of these failures may be referred to as mud flow. Such a flow occurs rapidly and the mass that moves may continue to flow to the lower ground a considerable distance away. Flow slides can take place in slopes as flat as 5 to 10°, and may result in slides of great magnitude.

Almost every stiff clay is weakened by a network of hairline cracks or "slickensides." If the spacing of the joints is wide, the slope may remain stable even on steep sloping ground during the dry season. However, if water is allowed to seep into the cracks, the shearing resistance of the weakened clay may become too small to counter the force of gravity and the slide occurs.

14.5 MAN-MADE SLOPES

Man-made slopes are necessary in the construction of highways, railroads, canals, and other projects. In high ground, open cuts with adequate slopes will be necessary and on low ground the stability of fill must be considered. Geotechnical engineers seldom pay attention to the design of man-made open cuts and fill. They cover the design with the standard construction specifications. For small projects with no seepage problems, such procedures may be adequate. However, for larger projects, such as locating a section of new highway, the designs of man-made slopes become critical. The cost of an over-designed slope can exceed the cost of a long-span bridge. At the same time, steep slopes can give maintenance crews years of headaches.

Experience has shown that slopes at 1 1/2 (horizontal) to 1 (vertical) are usually stable. The sides of most railroad and highway cuts less than 20 ft use such slopes. The standard slopes for water-carrying structures such as canals range between 2:1 and 3:1.

14.5.1 SLOPES IN SAND

Instead of failing on a circular surface, sand slopes fail by sliding parallel to the slope. Sand located permanently above the water table is considered stable and cuts can be made at standard angles. Slides occur only in loose saturated sands that liquefy.

When the slope angle exceeds the angle of internal friction of sand, the sand grains slide down the slope. The steepest slope that a sand can attain, therefore, is equal to the angle of internal friction of the sand. The angle of repose of sand as it forms a pile beneath a funnel from which it is poured is about the same as the angle of internal friction of the sand in a loose condition. Geotechnical engineers seldom have the opportunity to study the stability of a slope of sand, unless there is a sudden rise of the water table or after an earthquake.

14.5.2 Slopes in Clay

The stability of a slope of clay can be expressed by Coulomb's shearing strength and shearing resistance relationship.

$$s = c + \sigma \tan \phi$$

where s = shear strength, psf
c = cohesion, psf
σ = effective normal pressure, psf
ϕ = angle of internal friction

A chart of the stability number for different values of the slope angle β is shown in Figure 14.11. For homogenous soft clay with $\phi = 0$, the stability number depends only on the angle of the slope β and on the depth of the stratum.

If the value of ϕ is greater than about 3°, the failure surface is always a toe failure. The above figure can be used to determine the stability number for different

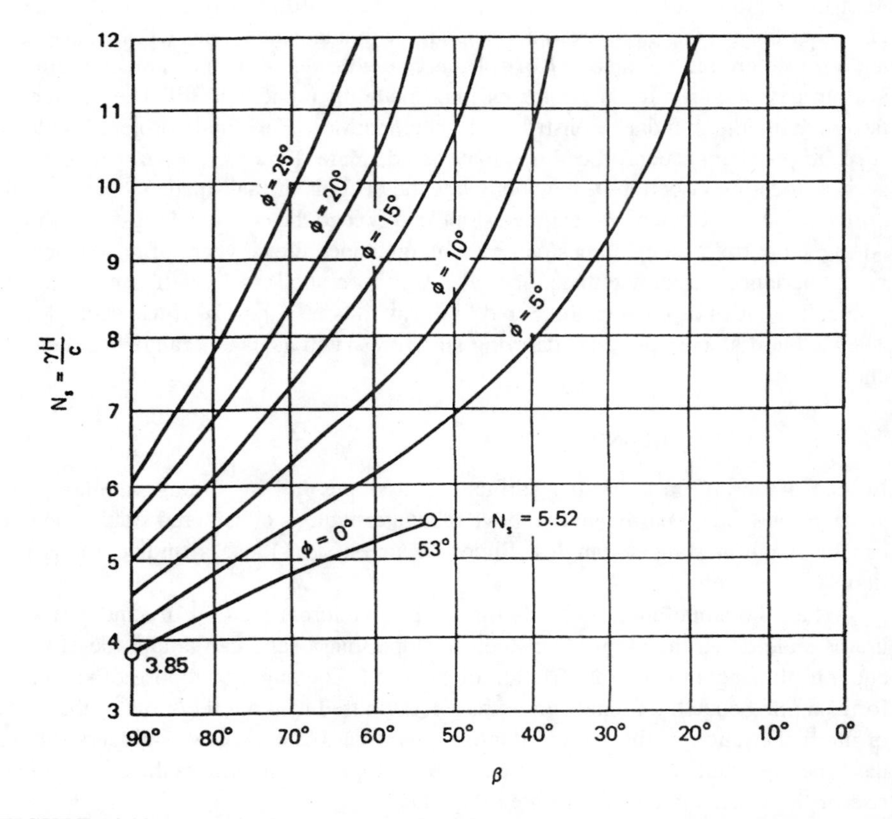

FIGURE 14.11 Chart for finding the stability of the slopes in homogenous unsaturated clays and similar soils (after Liu).

value, of ϕ, by entering along the abscissa at the value of β and moving upward to the line that indicates the ϕ angle and then to the left, where the stability number is read from the ordinate.

14.5.3 Slopes of Earth Dam

The design of a dam shell consists of the selection of the fill material on the basis of its strength and availability for construction. Generally, the material is obtained from borrow pits. Rock waste from a dam core excavation can also be used. For the upstream side of the dam, consideration should be given to the seepage problem. Thus, the degree of compaction of both soil and rock should be under control.

The slopes of most dams are established on the basis of experience and consideration of the foundation conditions, availability and properties of material, height of dam, and possibly other factors.

Typical upstream slopes range from 2.5(H) to 1(V) for gravel and sandy gravel to 3.5(H) to 1(V) for sandy silts. Typical downstream slopes for the same soils are 2(H) to 1(V) to 3(H) to 1(V).

A seepage analysis is made on the trial design to determine the flow net and the neutral stresses within the embankment and the foundation. Safety against seepage erosion and the amount of leakage through the dam are computed.

Stability analyses are made of both faces of the dam, using the method of slices. The upstream face is usually analyzed for the full reservoir, sudden drawdown, and the empty reservoir before filling. The downstream face is analyzed for the full reservoir and minimum tailwater and also for sudden drawdown of tailwater from maximum to minimum if that condition can develop.

The construction of the Greyrock Dam embankment at Greyrock, Wyoming is shown in Figure 14.12.

14.6 FACTOR OF SAFETY

The designs of cut and fill have been studied by many leading academicians. With the use of the computer, all aspects of dam design can be accomplished accurately and quickly. All studies involve some basic assumptions, as follows:

The soil is homogenous both in extent and in depth.
A single angle of internal friction and cohesion values can represent the entire soil mass under investigation.
The assumed seepage condition will remain unaltered.
There will not be any external disturbance affecting the stability.
The type of affiliated structure has not been determined.
There will not be any unexpected surcharge load.

In order to cover the above uncertainties, geotechnical engineers assign various values of "factor of safety" in an effort to cover the possibility of failures. Sower listed in Table 14.1 the factor of safety applied to the most critical combination of forces, loss of strength, and neutral stresses to which the structure will be subjected.

FIGURE 14.12 Embankment compaction, Greyrock Dam, Greyrock, Wyoming.

TABLE 14.1
Suggested Factor of Safety

Safety Factor	Significance
Less than 1.0	Unsafe
1.0–1.2	Questionable safety
1.3–1.4	Satisfactory for cuts, fills; questionable for dams
1.5 or more	Safe for dams

Sower further stated that under ordinary conditions of loading, an earth dam should have a minimum safety factor of 1.5. However, under extraordinary loading conditions, such as designing a super flood followed by a sudden drawdown, a minimum factor of safety of 1.2 to 1.25 is often considered adequate.

In addition to the uncertainty involved in the factor of safety, engineers must consider "cost"; with an exception of dam design, if cost is not a factor, the risk of a slope failure can be greatly reduced. The factor of safety values suggested by Sower should be considered as design guides.

The factor of safety against sliding is determined by dividing the sum of forces tending to resist sliding by the force tending to cause sliding. The slide-resisting forces are determined from laboratory testing on the so-called "representative samples." As stated in Chapter 5, the shearing resistance determined from the direct shear test or the triaxial shear test can be misleading. Consequently, the overall shearing resistance against sliding for a slope, as computed in the laboratory, can be very far from the realistic value. The term "factor of safety" as used by engineers must be qualified.

Unfortunately, the term "factor of safety" as understood by lawyers is quite different than that understood by engineers. When an engineer, in computation, indicates that the factor of safety is 1.5, attorneys consider the figure absolute. If upon further calculation the value is 1.4 instead of 1.5, a mistake is made, resulting in damage. The designing engineer must pay for the error. At the same time, if upon further calculation the value is 1.6, the engineer is clear and the responsibility for the damage must rest on someone else. Such logic, unfortunately, is agreed upon by judge and juries.

When dealing with soil, engineers should avoid the use of the term "factor of safety." Instead, the use of such language as, "We recommend ..." should be encouraged. By so doing, many of the legal problems involving geotechnical reporting can be minimized.

14.7 CASE EXAMPLES

14.7.1 HOWELSON HILL

Howelson Hill is located to the south of Steamboat Springs, Colorado. It is the site of international ski jumping competitions. In July 1976, a 41,000 cubic-yard landslide occurred during nearby excavation activity. The slide involved only surficial colluvial deposits that overlie the Morrison Shale formation on the slope of the north-south-trending bedrock.

The landslide area and all jump profiles were analyzed to determine the cause of the landslide and the effects the slide had on the profile of the jump complex design. A geological interpretation of all data was made. Stability analyses were then conducted using effective strength parameters as determined from laboratory tests. Circular and non-circular failure surfaces were considered in the stability analysis. Typical stability analyses are shown in Figure 14.13.

Correlation of the geology and strength parameters was made to assess the probability of similar slides in the adjacent jump area.

The following conclusions on the nature of the landslide were drawn, based on the field investigation:

1. The slide was localized in the area of previous instability and the geologic conditions were significantly different in the stable portions of the mountainsides.
2. The primary mode of failure is transitional.
3. A buried bedrock topography controls the lateral extent of the landslide, which is restricted to a narrow linear zone.

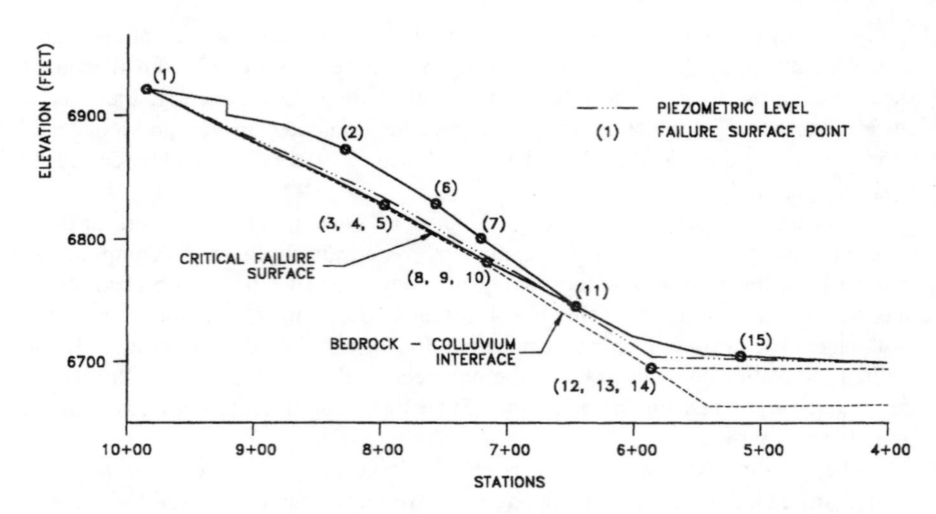

FIGURE 14.13 Non-circular analyses of final 70-meter profile.

4. Hot mineral springs were active at one time in the toe area of the landslide.
5. Older landslide movements have occurred in the area as indicated by the claystone shale overlying the colluvium in some test pits. Surface evidence of this older landslide is not present.

By the time the initial findings of the field investigation were determined, the rate of the movement of the slide had slowed down to less than 1 in. per day. The following remedial constructions were made:

1. A gravel trench drain was installed along the toe of the landslide.
2. A stabilizing berm of compacted soil was placed at the toe of the slide.
3. Open surface cracks in the slide mass were sealed by grading the slide surface.
4. Surface drainage ditches were constructed at the crown of the slide and at intervals along the slide.
5. To increase the factor of safety against sliding, retaining walls with foundations on the bedrock were constructed at the site of the 90-meter, 70-meter and 50-meter jumps.

After the completion of the remedial work, surface survey monuments were established and have periodically been surveyed. No measurable movement has been monitored.

14.7.2 Church Rock Uranium Mill Tailings Dam

The project site is located northwest of Gallup, New Mexico. The dam was to be 70 ft high and about 2 miles long. The purpose of the dam was to retain the waste from uranium mining at Church Rock. The site is in Pipeline Valley, that consists

FIGURE 14.14 Howelson Hill sky jump slope failure.

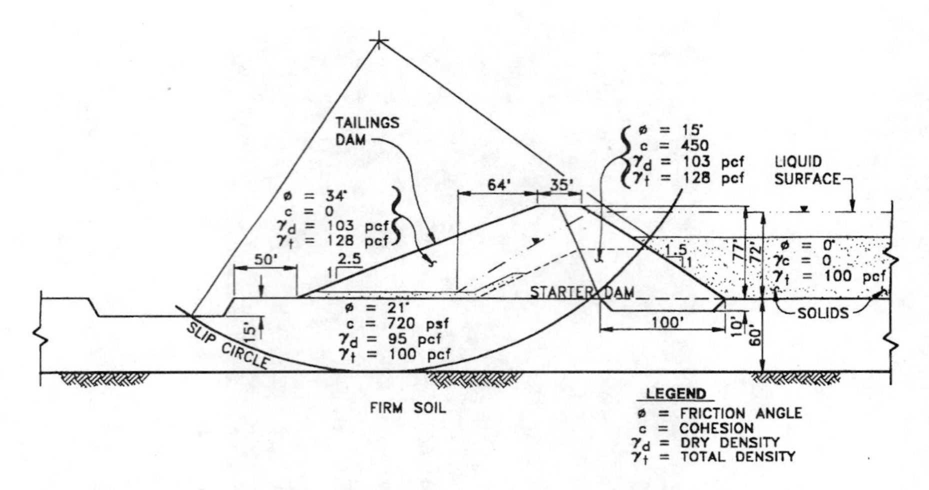

FIGURE 14.15 Church Rock Uranium Mill Tailings dam.

mainly of mancos shale and gullup sandstone, bounded to the north by Crevasses Canyon formation and to the south by the Morrison formation. The valley floor is topped with alluvium soil, its thickness varying from a few feet to more than 100 ft.

The design and stability analyses of the dam can be summarized as shown in Figure 14.15. The construction of the dam was to be carried out in stages. The height of the dam was to increase with the reservoir liquid level, until the maximum height of 70 ft was reached. Construction of the starter dam took place in 1976. The height of the dam reached about 30 ft when the accident took place.

Cracking of the dam embankment first appeared in June 1977 and was grouted. Further large separation cracks appeared at the south end in 1978. The cracks appeared to take place perpendicularly to the dam axis at the location where there is an abrupt change in the depth of bedrock as well as the thickness of the alluvium. Finally, in July 1979 the dam breached. All of the tailing liquid contained in the reservoir escaped through the breach.

A board of inquiry, consisting of many of the top geotechnical engineers in the country, was established to determine the cause of the failure. After months of investigation, the experts carefully evaluated all possibilities. They listed the following questionable items that may have caused the failure.

1. An excessively high free liquid surface, which came in contact with the transverse crack in the starter dam
2. The use of cyclones to form the coarse sand bench
3. The recommended freeboard of 5 ft was not maintained
4. Action between the liquid acidic tailing and the soil in the embankments
5. The foundation soil problem
6. Inadequate stability analysis or insufficient compaction
7. A low factor of safety, perhaps below 1.0
8. Piping in the embankment

9. Seepage through the embankment
10. Seismic effect

After careful evaluation, the general opinion was that the main cause of the failure was differential settlement. Prior to construction, it was predicted that the total settlement will be about 2.5 ft. Before breaching took place, the actual total settlement reached about 5 ft. Due to the extreme variation of depth to bedrock along the dam axis, differential settlement could be equal to total settlement.

Differential settlement resulted in embankment cracking. The introduction of reservoir liquid through the cracks caused soil erosion and resulted in piping. Other probable causes as listed above should not be ignored, but they are minor and are not considered the major cause of the dam failure.

14.7.3 Columbia River

Numerous landslides have occurred along the banks of the Columbia River in the vicinity of the Grand Coulee Dam, Washington. With the completion of the dam, the water level of the river rose and greatly affected the stability of the natural bank slope. The stability was further affected by the sudden drawdown of the reservoir. Stability of the bank slope was under intensive study by the Bureau of Reclamation. The study included the geology, the hydrology, and other related subjects. Publications of the studies are available in various technical journals.

In 1984, a staging area was developed immediately upstream from the Grand Coulee Dam. Contractors were allowed to place fill on the bank within the designated area. An accident occurred when a dump truck slid into the river, killing the operator. The accident stirred up not only legal responsibility but also the issue of stability of the existing slope along the banks of the Columbia River. Upon further intensive studies, the following subjects were brought up for review:
On the natural slope:

The stability of the natural bank slope
The effect of the water level in the river on slope stability
The effect of the sudden drawdown
The consistency of the soil property
The criteria used in the stability analysis
The factor of safety of the natural slope
The history of the bank slide
The geology of the staging area
The ground water level
The significant of the bedrock elevation
The existing slope 1.3:1 (38°); is that a safe slope?

On the new fill:

The maximum permissible thickness
The soil property of the fill
The design slope of the fill

FIGURE 14.16 Breaching of Church Rock Uranium Mill Tailings Dam.

The criteria used for stability analysis of the fill slope
The method of the determination of the soil constant
The factor of safety of the fill slope
The frequency of the soil tests
The extent of the fill placement
The berm establishment
The after slide slope 1.5:1 (33°); is that a safe slope?

On the combined bank soil:

The combined design slope
The combined factor of safety
The geometry of the combined slope
The interaction of the fill and the natural soil
The slump failure; is that a possibility?

On the victim's dumper:

The allowable weight of the dumper
The actual load carried by the dumper
The vibration of the dumper
The distance between the dumper and the edge of the berm

The above issues clearly indicated that the slope stability problem is much more complicated than the textbook analysis. In fact, the issues involved are so complicated that a rational solution cannot be readily found.

Finally, the case was narrowed down to the determination of the most desirable factor of safety of the embankment slope. All other pertinent engineering issues such as "effects of drawdown" were ignored. The case was settled out of the court. The cause of the accident was never determined.

REFERENCES

C.M. Hsieh, *Atlas of China,* McGraw-Hill, 1973.
C. Liu and J.B. Evett, *Soils and Foundations,* Prentice-Hall, Englewood Cliffs, NJ, 1981.
J.D. Nelson and E.G. Thompson, Creep Failure of Slopes in Clay and Clay Shale, 12th Annual Symposium for Soil Engineering and Engineering Geology, Boise, ID, 1974.
R. Peck, W. Hanson, and T.H. Thornburn, *Foundation Engineering,* John Wiley & Sons, 1974.
G.B. Sowers and G.S. Sowers, *Introductory Soil Mechanics and Foundations,* Collier-Macmillan, London, 1970.
D.W. Taylor, *Fundamentals of Soil Mechanics,* John Wiley & Sons, New York, 1948.
K. Terzaghi, R. Peck, and G. Mesri, *Soil Mechanics in Engineering Practice,* John Wiley-Interscience Publication, John Wiley & Sons, New York, 1996.

15 Distress Investigations

CONTENTS

In ancient times, structural damage was seldom a problem. Major structures were founded on selected stable grounds with uniform subsoil. Building material consisted mostly of timber capable of tolerating a great deal of movement without showing damage. When damage appeared, it was taken for granted and "nothing to be alarmed about."

In the last century, with the advancement of building technology, architects put steel, concrete, glass, and masonry into the same structure. Since each material has a different thermal expansion index, it is not difficult to expect that they cannot perform as a unit. Great architects such as Frank Lloyd Wright could not avoid structural damage. Structural engineers took most of the blame until the emergence of the foundation engineer. It was then – in the eyes of many – that every problem became related to foundation movement. Geotechnical engineers were helpless in trying to defend their designs.

Our society demands perfect performance from the building industry. It wants to pay the least yet it expects the most from investments. The legal profession is on the side of the consumers. The insurance business cannot afford to lose money so ιums skyrocket. Thus, a unique triangle has been established between the 'ofession, the legal partners, and the insurance business. The easiest culprits ιo blame are, of course, the design professionals.

When distress takes place, the owner wants to know the cause of the damage in order to establish the party responsible for the damage.

15.1 HISTORICAL

The first step toward an investigation of a distressed structure is to obtain its complete history. Unfortunately, such information is often lacking and it is necessary to uncover much of the required information by exploration.

15.1.1 Foundation Information

Efforts should be made to obtain information about the existing foundation. For buildings erected before 1960, such information is generally sketchy. Soil tests on individual sites have become a requirement since 1960. From the soil test data, it is possible to determine the following:

1. Type of foundation
2. Design criteria
3. Water table condition
4. Type of foundation soil
5. Moisture content of foundation soils
6. Settlement or swelling potential of foundation soils

Sometimes the subsoil investigation is not conducted for the specific building but for the general area. In such cases, the subsoil information has only limited use. Care should be taken to locate the nearest test hole to the structure under investigation, so that it is possible to determine as closely as possible the subsoil conditions beneath the distressed building.

The above information can be invaluable in finding the cause of movement.The second step is to check the foundation plan. Again, such information may not be available, either because the drawing is lost or the concerned party does not want to produce it. The foundation plan can reveal if the recommendations given in the soil report have been followed. These are:

1. The total load and the dead load exerted on the footings or piers
2. The size of footings or piers
3. The length of piers or piles
4. Pier reinforcement, pier connection, pier spacing, and others
5. Expansion joints, dowel bars, underslab gravel, slab reinforcement, and other details
6. Subdrain system
7. Backfill and lateral pressure

If the above information is available, the investigation can be greatly simplified. This is analogous to the case in which the complete medical record of a patient is at the disposal of the examining doctor. It is also necessary to examine the qualifications of the designer, including whether the design was made by a registered professional engineer or by a contractor.

When the above information is not available, it is necessary to expose the foundation system by excavation. In the case of a non-basement building, excavation can be easily performed outside the building immediately adjacent to the grade beam. In the case of a basement structure, it is necessary to break the concrete slab inside the basement in order to reach the foundation system.

For a deep foundation system, it is difficult to expose the entire length of the pier in order to determine the settlement or uplifting of the pier. An experienced engineer can sometimes gain a great deal of information from the examination of the upper portion of the pier.

Undersized footings can cause settlement problems. Oversized footings are the common cause of uplift in expansive soils. Misplaced footings can cause cracks induced from lateral pressure. Above all, if water is found seeping into the foundation system, it can be the source of all evils.

Logs kept by the driller are sometimes available. In such cases, a complete picture of the pier system will be helpful. This can also provide information on the depth of penetration of the piers into the bedrock. In case of a pile system, the penetration resistance and the blows of the pile hammer can be the key to the entire problem.

15.1.2 Movement Data

Efforts should be made to obtain the chronological data on the building's movement. Items such as the date the building was completed, the date the first owner moved in, and the date the first crack appeared should be noted in detail. All information obtained from the owner should be carefully scrutinized for its validity. If the owner intends to sue the builder to recover damages, the information may be exaggerated. With careful interrogation and keen observation, the true story can sometimes be revealed.

Since almost all foundation movement problems are associated with water, it is imperative that the investigation of the sources of water that entered the structure be the first order of the distress investigation. On the exterior of the building, it is helpful to find out the lawn watering practice, the setting of the automatic sprinkling system, and the condition of the backfill. Most owners deny the occurrence of excessive irrigation of the lawn and flower beds. They certainly will not tell you about the time they left the hose running for the entire weekend.

In the interior of the building, the primary information can be obtained in the basement area. Watermarks on the walls usually tell the story of seepage. A complete record on seepage water should be obtained, noting the first appearance of water in the basement, the location of seepage, the amount of observed water, and whether seepage has taken place after heavy precipitation.

Also important is the performance of the utility lines. Was plumbing difficulty experienced in past years? Had the floor drain been plugged? A recent investigation revealed that a house's sewer line was never connected to the main sewer line, and raw sewage accumulated around the house for as long as 15 years. It is fortunate that the house is still standing. In another case, the basement shower drain was not connected to the sewer line. For years, the error was not detected until the engineer entered the crawl space and found the excessive wetting condition. The obvious seepage problem is shown in Figure 15.1.

It is also important to know whether during construction there were any unusual incidents that contributed to the building's distress at a later date. If such information is available from an observant owner, it can unlock many movement puzzles. There are instances in which the soil was flooded during construction, and movement took place even before the building was completed. One case reported a partially completed house whose basement was covered with more than 2 ft of snow that the contractor failed to remove before enclosing the structure.

In the course of information gathering, all hearsay should be screened. Stories such as an underground river running under the structure, a building sliding downhill, and the bentonite in the soil pulling the building apart should not be believed. An experienced engineer will dismiss most of such hearsay and focus on concrete evidence.

15.2 INVESTIGATION

The cause of cracking of a distressed structure can usually be determined from visual examination by an experienced geotechnical engineer. However, at times it will be necessary to perform testing of the soil samples, either to obtain pertinent information or to confirm a preconceived opinion.

Owners sometimes feel cheated if the engineers did not perform drilling and sampling. They feel that by so doing, the cause of the problem can be revealed, not knowing that sampling the subsoil outside the structure can yield little information on the cause of distress. Most academicians prefer to have test data before them to support their findings. They may order drilling as the first step of the investigation.

15.2.1 TEST HOLES AND TEST PITS

Test holes should be drilled adjacent to the building, and sufficient samples should be taken for the determination of the consolidation or swelling characteristics and the moisture content of the soils. At least one test hole should be drilled away from the structure in an area unaffected by building construction. The physical characteristics of the soil obtained from the adjacent and remote test holes can be compared.

For a distressed structure, the soils immediately adjacent to the foundation generally have been wetted excessively. Laboratory testing on samples taken generally shows a low settlement or swelling tendency. Initial characteristics of the foundation soils can be revealed by air-drying the sample to the natural moisture content and then by subjecting it to wetting. Samples obtained from areas unaffected by building construction should give information relative to the soil behavior at the time of construction.

FIGURE 15.1 Water seeped into the lower floor.

The most positive method of determining the subsoil condition and construction details in the foundation system is by opening test pits. Test pits should be excavated adjacent to the grade beam and next to the interior supports. By investigating the subsoil in the test pit, the following can be revealed:

1. The foundation system
2. The moisture content of the soil; samples should be taken every 12 in. to a depth of at least 5 ft
3. The connection between the piers or piles with the grade beam
4. The presence of tension cracks on the pier shaft
5. The presence of underslab gravel, and the condition of the underslab soils

Distress investigation can be greatly simplified when crawl space construction is used in the system. By entering the crawl space, the engineer is able not only to inspect the foundation system but also to examine the drainage system, ventilation, pipe connections, surface soil moisture condition, and other details that the owner has ignored.

15.3 CAUSES OF DISTRESS

Most distressed structures that require investigation and remedial construction belong to a relatively low-cost category; the owner put in a great deal of money and expected a good return for the investment. The degree of the distress that appears in a structure varies greatly. Some of the distress is so severe that a complete demolition of the structure will be required. Others appear as hairline cracks which can be easily repaired. Some owners insist on complete and prompt repair and do not hesitate to take the matter to the court. After reviewing the design, the construction, and the maintenance practice, the foundation engineer should be able to provide a logical answer to the problem.

15.3.1 Foundation Design

Every state in the union now requires a foundation design with a professional engineer's seal. However, a soil report is not required for the foundation design. Some engineers are not familiar with the local conditions and adhere to traditional practices. An engineer not familiar with the swelling soils problems who designs the footings with settlement considerations can encounter serious problems. In selecting a foundation engineer, experience should be the first consideration.

Despite any deficiencies of a drilled pier or a footing construction, a properly engineered foundation system seldom suffers major distress even under the most adverse conditions. Some buildings that suffer severe damages are designed and constructed by contractors without the benefit of a soil or structural engineer.

Some contractors take the matter entirely into their own hands and prewet the foundation soil, puddle the backfill, reinforce the footings instead of the foundation walls, drill oversized piers, and expect a stable building. These buildings may remain

FIGURE 15.2 Timber supports used to correct the missed piers.

in good condition for years and may provide an excuse to the builder to continue this undesirable practice.

15.3.2 Construction

Legally, if the contractor follows every detail specified in the specifications and design provided by the engineers, his liability in the event of future structure damage can be cleared. Still, the contractor cannot avoid being named in a lawsuit. In court every effort is usually made to prove the contractor's negligence. For instance, the absence of slab reinforcement as specified should not be important with respect to slab cracking; however, in court, this can appear as a glaring mistake on the part of the contractor for not following the design.

There are cases in which the contractor abuses every rule of good construction practice and expects the infractions to go undetected. Faulty construction will not be uncovered until thorough inspections are made. Figure 15.2 shows the mislocation of piers.

15.3.3 Maintenance

Poor or neglected maintenance can be the cause of building distress. The most common poor maintenance practices are as follows:

1. Water ponding adjacent to the building
2. Flower beds located adjacent to the foundation walls
3. Tree roots extended under the building
4. Absence of outlet of the exterior drain

5. No outlet for the roof downspout
6. Flooded crawl space
7. Poor remodeling construction

Some owners purposefully aggravate the damage by heavy watering for the purpose of impressing the judge and jury in order to obtain better compensation.

If a civil action is instigated resulting from a foundation problem, the strongest defense the contractor possesses against the owner's suit is that the owner did not provide proper drainage around the building. An alert developer issues each home owner a manual on the care of drainage around the house. In the event of a complaint, the builder can then have recourse back to the owner.

15.3.4 Earthquake

Potential seismic activity is a major concern in structural design in many regions of the world. Earthquake potential must be considered in the design of major structures such as dams and high-rise buildings. Earthquakes can result in the direct tearing of structures that lie on a fault.

Whenever there is a light tremor, it gives a good excuse for all parties concerned to say that the cracking of the structure is an act of God. Indeed, it is difficult if not impossible to differentiate cracks resulting from earth tremor and cracks resulting from foundation movement.

One method of determination is to examine the mortar between the cracks. If the crack is due to an earthquake, the mortar is ground up into fine powder, while in the case of foundation movement, the mortar will remain in pieces.

15.4 STRUCTURAL MOVEMENT

It is easy to blame all building distress on foundation movement. Unfortunately, it is difficult to differentiate distress caused by foundation movement and distress caused from structural movement. In the Rocky Mountain area, most of the general public is fully aware of the swelling soil problem and believes that all problems are generated from swelling soils. Most structural engineers are aware that some movement can be caused from structural design, but hesitate to point it out. Why ask for trouble?

Structural movement problems are explained well in Richard Weingardt's paper "All Structures Move." He stated: "The structures are like people, they breathe." The following is part of Weingardt's analysis.

15.4.1 Deflection

The normal allowable deflection, by most codes, is 1/360. This translates into about a 5/8 in. deflection in a 20-ft span. Deflections of this magnitude are not uncommon. Excessive deflections may result in curvatures or misalignments perceptible to the eye. Lesser deflections may result in fractures of elements, such as plaster or masonry.

Deflection of load-carrying members may result in the transfer of loads to non-structural elements, such as window frames or interior partitions. This situation must often be handled by detailing connections that allow for a limited movement. This requires close coordination of the architect and the structural engineer.

15.4.2 Thermal Movement

Exterior elements and surfaces of buildings are subjected to a range of temperatures which in northern climates can be as much as 140°F. The resulting expansion and contraction can cause cracks similiar to cracks from foundation movement, as shown in Figure 15.3.

There is movement of the entire building structure. Since the substructure is kept at a relatively constant temperature by the ground, there is a buildup of differential length between the lower and upper portions of the building, which can cause critical conditions of shear in end-walls or sideways deflection of end columns. Where separate wings of the building are joined, as in L- or H-shaped plans, one wing may pull away or push against the other. The solution to this problem is usually to provide expansion joints.

FIGURE 15.3 Cracks due to temperature in massive brick construction.

Different materials have different thermal properties. When exposed to temperature changes, they should not be rigidly connected to one another. Steel beams or columns encased in masonry should be provided with a void or slip jointed ends.

Another recommended standard detailing practice is to weld alternate ends of steel joists or prestressed members, so they are allowed minor movement without tearing themselves apart.

15.4.3 Prestressing

The creep of prestressed concrete members such as twin-tee floors and roofs can continue for many years. Figure 15.4 indicates a severe crack that developed in a twin-tee structural slab. The building is founded with drilled piers, and the exterior walls are in excellent condition. These cracks had nothing to do with the foundation movement.

It is easy to jump to the conclusion that the foundation system has moved and is responsible for the structural damage. It is sometimes impossible to prove that the damage is actually related to structural movement in the other component parts of the structure.

Weingardt, in the conclusion of his paper, stated: "The awareness that structures move, in itself, is the first step in designing buildings, to eliminate cracking. There is not one simple solution for an expansion joint or control joint that will work for all materials. Masonry walls are different from concrete walls. Concrete slabs are different than steel beams."

FIGURE 15.4 Floor crack due to creep of twin-tee concrete slab.

15.5 DISTRESS OF MAJOR STRUCTURES

When the distress or failure of a major structure takes place, it may result in litigation. The sequence of events usually starts with the owner concerned about the integrity of the structure. Owners blame faulty construction. The contractors blame everyone concerned except themselves. The owners turn to academicians for expert opinions. The professors, who usually have little experience in dealing with foundation movement, use various theories, such as plastic flow, wave propagation, diffusion, etc., to explain the causes of movement. None can confirm or dispute their theories. It is of course easy to discuss the design after distress has taken place. Bear in mind, when the geotechnical engineers take on the assignment to recommend the design of the foundation system, they are not paid to write a thesis on the project.

It takes months and sometimes years to complete the depositions in a lawsuit. Expert witnesses may include professors, builders, surveyors, architects, or insurance agents; the experienced geotechnical engineers are seldom involved. Ninety percent of the cases are settled out of court. The defendants would rather forget the entire matter, than spend additional money to fight it in court. For the limited cases that are tried in court, the issues are usually badly distorted. The attorneys tend to present some unimportant events and elaborate on them. The juries are at a complete loss listening to the theoretical approaches advanced by the academicians. The truth of the matter is seldom if ever revealed. The winners of the entire episode are the attorneys.

15.5.1 DRILLED PIER FOUNDATIONS

Drilled piers are used for most of the major structures in the western U.S. Distress of structures caused by improper design of the pier system is seldom reported.

One example cites that during the construction of a warehouse building, several piers had settled more than 3 in. The tilt-up panels were jacked up and shimmed, but settlement continued. After intensive study, it was concluded that the problem resulted from the improper placement of pier concrete. A stress wave propagation testing method was introduced to determine the condition of the distressed piers.

Stress wave propagation testing consists of inducing a shock wave at the top of a pier and measuring the compression wave propagation and attenuation characteristics with an oscilloscope that is connected to a receiver mounted on the top of the distressed pier. Test results can be used to identify voids in concrete, voids at the base of a pier, and disturbed conditions along the piers.

Test results indicated the piers that suffered excessive settlement exhibited higher reflection numbers, thus indicating the presence of relatively softer materials surrounding the piers.

Stress wave propagation tests were followed by load tests. The selection of the piers for load tests was based on the results of wave propagation tests. The piers were loaded to approximately 60 kips by means of a 50-ton jack reacting against a crane. The results of the load tests are shown in Figure 15.5.

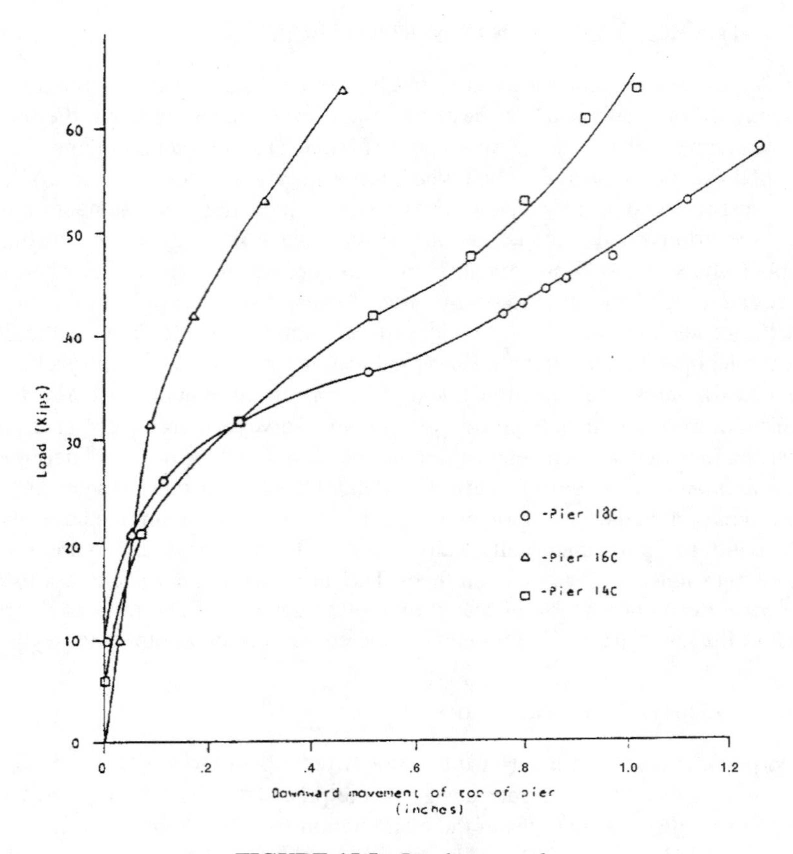

FIGURE 15.5 Load test results.

Tests indicate that piers 16-C and 14-C were both deformed more than generally acceptable at the applied load. It was then decided to excavate adjacent to the distressed piers to evaluate the soil and rock conditions along the sides and beneath the piers. Excavation revealed that a significant portion of the zone adjacent to the sides of the piers contained disturbed materials. The material was about 1 in. thick and had a high moisture content relative to the natural on-site materials.

It was concluded that the cause of the excessive settlement was from the pier's installation procedure. It was possible that the caisson rig used did not have the auger properly attached to the kelly bar with the axis of the coincident with the axis of the kelly bar. It was also possible that the auger was either loosely attached to the kelly bar allowing it to wobble, or there was a bend in the auger or the kelly bar.

The defective piers were replaced and construction continued with no further problems. Since the pier drilling operation was inspected by the geotechnical consultant, the cost of the investigation and remedial construction was shared by the consultant and the contractor. Mishaps similar to this case can hardly be avoided even with careful inspection.

15.5.2 Watering Practice

A nursing home complex with a drilled pier foundation is located in a swelling soil area in Colorado. About 2 years after the completion of the buildings, the floor slab in a part of the building heaved. Floor movement resulted in the cracking of the slab-bearing partition walls and the distortion of the ceiling structure. An open courtyard was located at the central portion of the building complex. The courtyard was covered with a lawn and flower beds and was heavily irrigated. The design team advised the administration to regulate its watering practice but was ignored.

The case was brought to court. After 2 years of intense litigation, the administration finally obtained an expert from an eastern university who had no experience in expansive soil. After examining the distressed photographs, the expert recommended the demolition of all buildings.

Geotechnical engineers finally convinced the owner to convert the lawn-covered court to a concrete playground and regulate the watering practice around the buildings. The building's movement was arrested and the necessary repairs made. Today, the buildings are in such good condition that the owner decided to build additions using the original design criteria.

The cost of this litigation far exceeded the total cost of the buildings. The case illustrates the danger of selecting the wrong expert in our legal system. It is a shame that some of our professionals are willing to become advocation for an attorney.

15.5.3 Building Demolished

An unfortunate case of distress investigation took place in Manoa, Hawaii. A 5-story concrete and masonry structure was demolished only a few years after completion. Both the foundation system and structural analysis of the building were made by leading nationally renowned consulting engineering firms. The distressed structure was a part of the building complex that varied from two to six stories in height and was connected by corridors at the upper level. The outstanding features of the initial investigation are as follows:

1. A soil report was made for the entire building complex. No test hole was made at the site of the distressed building.
2. The upper soils consist of stiff, highly plastic silty clays with varying amounts of decomposed basalt gravel.
3. A spread footing foundation placed on the upper soils designed for a bearing pressure of 5000 psf was recommended.
4. It is likely that fill was placed on the site. No control of fill placement was made.

The cracking experienced at the building was essentially in the horizontal pattern and major cracking took place at the header beam with the column. Vertical cracks occurred in the brick course, but only a few with a diagonal pattern.

The cause of cracking was investigated by fewer than a dozen experts in geotechnical as well as structural engineering. Since some tests indicated that the foundation soils expanded as much as 10%, it was generally concluded that the cause of cracking was due to the swelling of the foundation soils. The conclusion was questionable due to the following analysis:

1. The natural moisture content of the foundation soils exceeded 50%. At this high moisture, content swelling could not have taken place.
2. The confining pressure used for all the swell tests was very low, on the order of 200 psf. Under the design load of 5000 psf, there was no chance that swelling could take place.
3. The pattern of cracks was not that of the typical swelling cracks.

Finally, the authority decided to demolish the building, with the following statement:

> "The damage is so severe and extensive that repairs cannot be economically made. The most economical and only positive remedial measure is to demolish the entire structure, including the foundation, and build new."

The cause of distress of this building was never determined. It was suspected that differential settlement had taken place between the foundation placed on the natural soil and that placed on the poorly compacted fill. The extent of distress experienced at this building certainly did not justify the demolition of this structure. Needless to say, the cost of this undertaking was shared by the geotechnical engineer, the structure engineer, and the contractor.

15.5.4 Debris Flow

The Telluride Colorado Airport runway is underlain by Dakota sandstone, Mancos shale, and glacial drift. The airport runway is 75 ft wide and about 7700 ft long. In connection with runway construction, a fill approximately 140 ft deep was extended across a ravine tributary to Deep Creek near the middle of the runway. Another fill having a maximum depth of 70 ft was placed across a small area near the end of the airport runway, as shown in Figure 15.6.

A portion of this fill slumped before the runway was completed, during a major debris flow event. The amount of earth movement was estimated around 70,000 cubic yards. The event disrupted the runway construction, and a lawsuit was brought by the Regional Airport Authority against the contractor and the engineers. The damage and the remedial construction was estimated at over $5 million.

The investigation into the cause of the accident and the responsibility was undertaken by eight geotechnical engineers, four professional geologists, and two civil engineers. Total cost of expert witnesses' expenses exceeded $140,000.

Experts were not in agreement as to the cause of the landslide or debris flow. According to one school, the possible causes are as follows:

1. The placement of fill interfered with the ground water regime, blocked drainage, increased the pore pressures in the soils, and ultimately caused the failure.

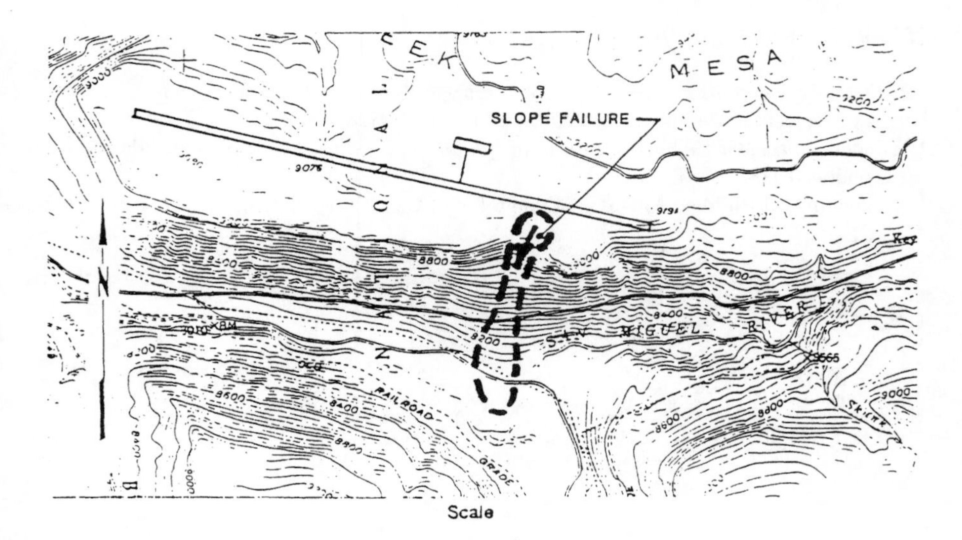

FIGURE 15.6 Location of debris flow.

2. The slide was caused by the placement of the uncontrolled fill. There was not adequate compaction, moisture control, or benching of the uncontrolled fill into the controlled fill. Differential settlement of the fills, which created vertical cracks, allowed water into the fill and ultimately caused the failure.
3. The rise of ground water after construction, and spring runoff, as well as the inability of the drainage to handle the increase of water, contributed to the instability.
4. The recommendations of the initial soil report were not completely followed.
5. The engineer failed to meet the standard of care for the design of the foundations for the embankments.

Another school about the possible cause of the failure is as follows:

1. The failure was due to a naturally occurring debris flow triggered by excess water from snow melt; it would have happened even if the runway fill had not been in place.
2. The drainage system constructed under the runway fill had no significant influence on the failure.
3. It is unlikely that an engineering geologist, even one experienced in debris flow analysis, would have identified the imminent flow failure zone as a high-risk geological hazard.

The case was settled out of court. The true cause of failure was never determined. Failure investigation of this case illustrates the merit of joint study between geotechnical and geological engineers. It also indicates that soil mechanics is an art.

REFERENCES

F.H. Chen, *Foundations on Expansive Soils,* Elsevier Science, New York, 1988.

Effects of Defects in Composite Materials, STP 836, ASTM, 1984.

P. Rainger, *Movement Control in Fabric of Buildings*, Batsford Academic and Educational, London, 1983.

R. Weingardt, All Building Moves — Design for It, Consulting Engineers, New York, 1984.

16 Construction

CONTENTS

The final product of an engineering project is a combined effort of the architect, the engineers, and the contractor. For an outstanding project, the architects get the honor and credit and will be remembered. We all recognize the famous architect, Frank Lloyd Wright, yet the builder is seldom mentioned. At the same time, if there is problem in the project the first one to blame is the contractor. It is the contractor who uses substandard material, abuses the specification, ignores the advice of the consultant, and causes the problems.

When damage appears in a structure, foundation failure is at once suspected. The contractor must prove that the construction complies in every respect with the plan and specifications. The contractor must produce evidence that foundation quality control has been followed. Quality control begins with the geotechnical investigation of the site and continues through the construction with proper construction control.

The relation between contractor and geotechnical engineer can be illustrated by Irving Youger's reply to a plaintiff when his cabbages were eaten by the defendant's goat.

> You did not have any cabbages,
> If you did, they were not eaten.
> If they were eaten, it was not by a goat.
> If they were eaten by a goat, it was not my goat.
> And if it was my goat, he was insane.

Ultimately translated to a foundation failure case, the defense might be postulated as:

The contractor was not negligent.
If the contractor were negligent, the geotechnical engineer was comparatively negligent.
Even if the contractor were negligent, the negligence did not cause the failure.

The actual theories are more than a little like the crazy goat reference above.

16.1 SCOPE OF CONTRACTOR

In the beginning, all the parties involved in the design and construction of a project must share the success or the failure of a project. The owner hires the architect. The architect in turn hires a variety of engineering services, including the geotechnical engineer. The architect compiles all the information and makes a recommendation to the owner concerning the design and construction of the project.

After the design is accepted, the general contractor is selected to construct the project. The architect and the owner generally share in the responsibility of the selection since both have a vested interest in choosing a qualified contractor capable of completing the work.

In the case of geotechnical engineering, a quality assurance program is generally accepted as part of construction. This program can include periodic or full-time observation and testing as well as documentation of adherence to the recommendations set forth in the soil report.

In the case of a subdivision development, the developer sometimes acts as both owner and contractor. The developer hires an in-house architect and sometimes an in-house engineer. Most of the time, the only other engineer he seeks is the geotechnical consultant. When the owners of the residence sue the developer, the liability will be entirely on the contractor and sometimes the geotechnical engineer.

16.1.1 SUBCONTRACTOR

The general contractors subcontract the project to the following specialized firms:

Earth Moving — For site preparation, fill removal, and fill placement. The testing and observation of this operation is generally left in the hands of the geotechnical engineer. The placement of local fill, such as in a narrow ditch or backfill, is usually placed without control.

Pier Drilling — The constructions of drilled pier foundation systems are usually subcontracted to the specialized firms. The general contractors seldom interfere with the pier drilling operation, except when there is an overrun or when unexpected problems are encountered.

Pile Driving — The performance of the pile-driving operation is observed by the geotechnical engineer. The general contractor is concerned about the time schedule and cost overrun.

16.1.2 Owner's Responsibility

A contractor may have faithfully performed all the obligations contained in the contract, including those of the subcontractors — yet the building suffered damage after occupation. Should the contractor be held responsible?

Legally, although the project has been accepted by the owner, that does not relieve the contractor of responsibility. Many buildings have suffered damage long after occupation, and the contractor as well as the engineers and architect are sued by the owner. The arguments by the defendants are:

The contractor cannot control the maintenance of the building, and if the owner chose to abuse the building, the responsibility should lie with the owner.

The engineer, especially the geotechnical engineer, can only give recommendations but not insurance; it is up to the insurance company to guarantee the integrity of the structure.

The most common maintenance negligence is described in Chapter 15.

As an example: a municipal building is founded with drilled pier foundations in an expansive soil area. The front part of the building suffered damage from pier uplifting. The owner sues the contractor for poor pier placement and the geotectnical engineer for assigning insufficient dead load on the pier. After many months of investigation, it is suspected that water from excessive lawn watering caused the pier uplift. The owner argues against the allegation.

It is decided that the owner and the geotechnical engineer should inspect the subsoil conditions under the crawl space. Upon entering the limited entrance into the crawl space, to the surprise of the owner the ground at the front portion in the crawl space was flooded. The owner agrees to drop all claims.

Concrete slabs, placed directly on the ground, are much less expensive than structure slabs with no direct contact between slab and ground. In an expansive soil area, geotechnical engineers recommend the use of structural slab, unless the owner assumes the risk of floor movement. In such a case, all slab-bearing partition walls should be provided with a void at the bottom, so that uplifting of the slab will not affect the upper structures. The engineer provides the contractor with detailed plans.

At a later date, the owner decides to add partitions in the basement. The partitions are placed directly on the floor with no void space underneath. Consequently, the lifting of the partition walls causes severe distortion of the upper structure. The owner blames the contractor for such damage. The geotechnical engineer is able to determine that the fault lay with the owner and not the contractor.

16.2 CONTRACT AND SPECIFICATION

The American Institute of Architects publishes "General Conditions of the Contract for Construction" for use by the architect. The contents of this document increase each year; it is now over 20 pages long in closely typed form. With this document, the owner and the architect sign the "Standard Form of Agreement Between Owner and Architect."

Before signing the contract, the architect will provide the contractor with all designs and drawings concerning the project. These include a soil report with logs of exploratory holes. Most construction contract bid documents require a site examination before the contractor bids on the project. In such cases, the contractor must make a prebid examination as a prerequisite to recovery for a changed condition. This duty of the bidder must involve a site inspection, but generally would not require an independent boring or test pit investigation.

However, if the contractor does not agree with the subsoil conditions depicted by the soil engineer, he should at his own expense conduct a separate subsoil investigation. In almost every case, the contractor accepts the soil report provided by the owner.

Construction is a risky business. Between signing of the construction contract and the ultimate completion, a number of factors invariably impact the project. The contractor assesses the risks of such factors and includes a contingency factor in the bid to protect himself. In such cases where the unanticipated subsoil conditions do not materialize, the contingency factor increases the contractor's profit. If the risk actually encountered exceeds those assumed, the contractor will generally seek to obtain relief by the filing of claims.

In order to reduce the number of disputes and claims, owners have included in their construction contracts provisions for risk-sharing by the owner and the contractor. These risk-sharing clauses include "Changes," "Differing Site Conditions," and "Suspension of Works," among others.

16.2.1 Differing Site Conditions

The contractor shall promptly, and before such conditions are discovered, notify the contracting officer in writing of subsoil conditions at the site differing materially from those indicated in the contract.

Most contracts allow two types of differing site conditions. Type I applies where the conditions actually encountered differ from the conditions indicated on the contract documents. This includes the following:

Groundwater
Rock and boulder
Bedrock
Man-made objects
Utilities
Soils

Discussions on Type I differing site conditions are given in Chapter 17.

The Type II differing site conditions are those conditions which are unknown and unusual when compared with conditions customarily encountered in the particular type of work. This category of differing site conditions is less frequently alleged and considerably more difficult to prove because the contractor must prove he has encountered something materially different from the "known" and the usual.

In one case, the site for the building of a bridge was flooded because of a diversion of the river by another contractor who was constructing a highway upstream. The court held that the diversion was a changed condition even though it had occurred after contract award.

16.2.2 Specifications

Standard general conditions of the construction contract are available from the following:

National Society of Professional Engineers
American Consulting Engineers Council
American Society of Civil Engineers
Construction Specification Institute

The above contract and specification forms can only be used as guides. For each project, a new specification should be drawn. Many architectural offices consider the preparation of specifications part of a routine procedure and tend to copy specifications used from the previous job with little modification. The contractors sign the contract without carefully reviewing the specifications.

Basically, where specifications prove impossible or impractical in terms of achievement, the owner is held liable for the cost incurred by the contractor in trying to perform as specified. Fill placement is the portion of specifications that has caused the most disputes. Defective specification can include the following:

Compaction equipment
Moisture content
Number of passes
Degree of compaction (standard or modified)
Thickness of each lift

As an example, the specifications require that the fill be compacted to 100% modified Proctor density, where in the soil report the required density is 95% standard Proctor density. The difference between modified and standard Proctor density that may be obvious to a geotechnical engineer can cause confusion to the architect or the contractor. The specification requirement and soil report requirement can lead to considerable change in the cost of site preparation.

16.3 FOUNDATION CONSTRUCTION

Foundation construction constitutes an important portion of a project. For a site with difficult subsoil conditions such as high ground water, pockets of fill, expansive or soft soil, etc., the cost of foundation preparation can be as high as 50% of the total project cost. Yet both the owner and the general contractor pay more attention to the above-ground work than the below-ground construction.

In the past, construction claims on foundation work were relatively small and frequently resolved at the field level. Nowadays, claims that arise may total a considerable percentage of the project cost and are rarely quickly resolved at the field level. Over the years, earthwork and foundation construction claims have become more sophisticated in terms of technical arguments and cause and effect relationships; these are often the prime issues at dispute.

Problems with footing foundations are generally limited to conditions of the under footing support. The presence of foreign materials or the loose and uncompacted soils beneath the concrete footing pads can easily be corrected.

At the same time, defective piers or piles can constitute a major problem, and the defective elements may not be detected until after the superstructure is partially completed.

16.3.1 Drilled Pier Foundation

It is this part of operation that gives the general contractor the greatest concern. Defective piers discovered after the building is partly completed can deal the general contractor a substantial financial loss. Legal involvement can be prolonged, delaying the completion and sometimes resulting in abandonment of the project.

Small-diameter piers on the order of 12 in. constitute the major portion of residential foundation systems in the Rocky Mountain area. Such piers generally are drilled without any supervision. Thousands of such piers are drilled each day with minimal complaints.

Large-diameter piers, in excess of 72 in. and more than 100-ft long, are used to support major structures in the western U.S. Such piers must be handled by experienced drillers under close supervision. Typical specifications for large-diameter drilled piers are as follows:

Pier Embedment — The project plans are indicative of subsoil conditions and depths where satisfactory bearing material may be encountered. If satisfactory material is not encountered at plan elevation, the bottom of any drill hole may be lowered. Alteration of plan depth will be made to comply with design requirements. Raising of the foundation elevation shall be approved by the engineer. If the drilling operation reaches a point where caving conditions are encountered, no further drilling will be allowed until a construction method is employed that will prevent excessive caving. If steel casing is proposed, the shell shall be cleaned and shall extend to the top of the drilled shaft excavation.

Cleaning of pier holes — After the completion of the drilled shaft excavation and prior to the placement of the reinforcing steel cage and concrete, all sloughage and other loose material shall be machine-cleaned from the shaft. A continuous flight auger or other equipment shall be used for cleaning dry excavation where slurry or ground water is not present. Where slurry or ground water is present, the excavation shall be cleaned with a bucket auger or similar type of equipment, as approved by the engineer.

Reinforcing Steel — The reinforcing steel cage for the drilled shaft consisting of longitudinal bars and spiral hooping shall be completely assembled and placed into the shaft as a unit. Spacers shall be inserted at sufficient intervals along the shaft to ensure concentric spacing for the entire length of shaft.

Concrete Placement — Concrete shall be placed as soon as possible after completion of excavation of the drilled shaft and the reinforcing steel placement. Concrete placement shall be continuous in the shaft to the top elevation shown on the project plans. For placement in dry excavations, concrete shall be placed through a tremie to prevent segregation of material.

Casing Removal — During removal of any casing, a sufficient head of not less than 5 ft of fluid concrete shall be maintained above the bottom of the casing. If any upward movement of the concrete or reinforcing steel occurring at anytime during the pulling operation is greater than 1 in., the casing shall be left in place as a permanent sleeve at the contractor's expense.

Other details such as pier mushroom, pier plumbness, concrete vibration, time lagging, delayed operation, abandoned pier holes, etc., may or may not be included in the specifications. The engineer is to determine the adequacy and acceptability of the drilled shaft.

16.3.2 Driven Pile Foundation

Specifications for pile foundations are more complicated than those for pier foundations. Instead of specifying a generalized condition, it is necessary to study each project separately. The following case illustrates the difference between the specified condition and the actual working condition:

Specification — The final pile design required the contractor to pre-drill through the compacted fill and gravel into the top of the hard clay stratum.

The piles used in the design were 12 × 12 in. steel H-piles with 50-ton design capacity. The piles had to be driven by a steam or air-operated hammer developing at least 32,500 ft/lb of energy per blow.

Piles had to be driven to an elevation –50 ft unless a resistance of 15 blows per in. was recorded before the tip elevation had been reached. If there was any difficulty attaining the specified tip elevation, the engineer was to be immediately informed and an alternate procedure was to be adopted upon his recommendation.

Construction phase — The contractor wrote to the owner requesting that pre-drilling be eliminated. The geotechnical engineer agreed to the change but the responsibility for driving the piles without pre-drilling has to be left to the piling subcontractor.

A test pile was driven with 32,500 ft/lb of energy per blow to tip elevation –51 ft and final blow count 200 blows per foot (16.7 blows per in.). The pile was loaded to 150 tons and settled 0.5 in. It was subsequently loaded to failure at 280 tons.

The geotechnical engineer changed the pile specification to raise the pile tip elevation from –50 feet to –43 feet and in certain areas allowed it to stop at elevation –37 feet.

As suggested by the geotechnical engineer, pre-drilling with auger was conducted in an effort to facilitate the driving effort. The hard driving situation continued.

A very heavy hammer rated 44,000 ft/lb of energy was brought in. The hammer was much heavier than the specified hammer. The job was eventually completed.

The contractor claimed that the 45 days' delay in completion of the pile-driving contract caused a 66-day delay in the project completion. The change of condition claim exceeded the initial contract amount.

The above case indicates the importance of a thorough geotechnical investigation, necessary before a specification is prepared and before the contract is drawn up.

16.4 CONSTRUCTION CONTROL

General and detailed specifications and drawings describe the conditions that must be met by the contractor. They are the technical basis for all matters regarding the project, and the contractor is obligated to follow all stated requirements. Any changes to the plans and specifications must be authorized by the designers.

The foundation phase of the project is generally covered up. The adequacy of the performance must be determined during the construction. Therefore, most contracts provide a section on "Field Inspection." The owner asks the geotechnical consultant to provide an "inspector" to control the performance of the contractor.

Contract wording is crucial. To the lay person, the distinction between on-site inspection, supervision, and observation may seem slight, but selecting the correct wording for your contract could determine the outcome of litigation. As a general rule, the architect or engineer on a project does not undertake the day-to-day overseeing of construction activity; rather, they are responsible for periodic observation.

Many contracts nonetheless call for the architect or engineer to supervise during the foundation construction phase, and courts have stated that the duty of supervision goes far beyond what is normally envisioned by the architect when entering into a contract. It is the engineer's responsibility to ensure that the contract correctly reflects the scope of service the consultant anticipates performing.

Today we understand that "inspect" is a dirty word. Attorneys argue that to "inspect" implies "warrantee." Therefore, the geotechnical consultant must be responsible for the performance of the contractor. We now elect to use such words as "observe," "examine," "oversee," or "supervise." No matter what word we choose to use, lawyers may find loopholes where they are able to drag the field person into the dispute.

16.4.1 FIELD TESTING

In most circumstances, testing is done with a nuclear moisture-density gauge. Sand cones and balloon methods are less commonly used. The sand cones are sometimes utilized to correlate the accuracy of the nuclear gauge, as described in Chapter 9.

Testing frequency is often defined in the specifications. Any departure from the testing frequency must be with the consent of the architect. If testing frequency is not specified, a sufficient number of tests should be taken to demonstrate the density and moisture content to each layer of fill. A general rule of thumb is to take at least one test for each lift per 1000 ft^2 in structural fill; one test per 500 cubic yards in overlot fill; and one test per 300 linear ft of roadway. More tests should be taken at the start of the project to establish a satisfactory construction procedure. A minimum of three tests should be performed on each visit of the technician. If compaction problems are encountered, additional tests should be taken to isolate the problem area. Failures must be retested unless the technician is able to verify that the unsatisfactory areas have been removed. Retests should be recorded on the daily observation report.

16.4.2 Moisture Control

Prior to placing fill on a natural ground surface, all old fill, topsoil, and vegetation should be removed. The area should then be scarified, moistened if necessary, and compacted as specified in the plans and specifications. This is done to provide a uniform base for subsequent fill placement.

While being compacted, the material in each layer should be at or near optimum moisture. Moisture should be kept uniform throughout the layers at both the surface and through the depth of the fill. This may require watering the material at the borrow area, watering during the placement, or mixing water with the fill after placement. The technician should follow the moisture requirements outlined in the plans and specifications.

16.4.3 Record-Keeping

It is important that all correspondence between the geotechnical engineer and the architect, the owner, the strucural engineer, the contractor, and any other individuals related to the project be kept. Even a note or a reminder that initially appears to be of no consquence may turn out to be an important document in a court of law.

A geotechnical consulting firm was sued by the government for failure to point out what was specified in the defective piers during inspection and for changing the size of the pier from the plan. Upon research on several-year-old records, the firm found the following document:

> "... the owner shall employ a soil engineer to inspect the bearing material, and the piers shall be inspected by the architect immediately prior to placing of concrete..."

Also a note from the structural engineer that reads:

> "... the contractor had asked for a substitution to use 10-in. round piers instead of 8-inch. Upon checking the structure engineer approved the change..."

These two records clear the geotechnical engineer of a very serious responsibility. The importance of record-keeping is thus obvious. Some even claim that the records should be kept as long as 10 years after the completion of a project.

16.5 THE TECHNICIAN

The technician is the consultant's representative. The technician's primary role is to see that the work conforms to the intent of the plans and specifications. The technician is primarily interested in results rather than methods, but because methods affect results, he or she must be able to detect improper procedures and suggest more appropriate ones. The technician also has the obligation to keep the job moving smoothly and avoid interference with the construction schedule. The observations made by the technicians call for tact and judgment. Technicians provide advice and make recommendations to the owner. If defects are found in the construction, it is the technician's duty to notify the owner so that the error can be corrected. As part of that duty, the technicians keep the contractor apprised of his or her advice to the owner. Understanding of the contractor's problems will ensure a cooperative relationship between the contractor and the technician, but the technician's primary responsibility is to the client.

16.5.1 Qualifications of a Technician

To be a technician, one must be tactful, honest, restrained, firm, alert, and patient. He or she must be knowledgeable about construction procedures, material, and equipment, and use good judgment based on experience, education, and training.

A technician may have risen from the ranks of the construction workmen and, by implication, have little or no academic training. All technicians should be acquainted with common construction practices, understand project specifications, and be able to interpret contracts and drawings. The technician must be able to make reports and maintain a diary of his observations that includes any warnings and instructions given to the contractor. Such records may turn out to be part of defendant's argument in a court of law.

The contractor's foreman with years of experience can usually tell if the technician has previous experience. With an inexperienced technician, the foreman might well be able to short-cut any operation without the knowledge of the technician.

16.5.2 Length of Service

Geotechnical services are provided on a full- to part-time basis, as determined by the person who fills out the job order. On full-time projects, the technician is responsible for observation of all work performed. On part-time projects, technicians must know the extent of the work being performed since their last visit in order to perform the appropriate tests. The technician should be satisfied that the work completed in his or her absence is satisfactory and watch for errors and deficiencies made during that time. If he or she has any doubt, additional testing should be performed to verify the extent of any deficiencies.

In order to save costs, most owners only allow part-time inspections. Part-time or the so-called "on call" service can be very dangerous. A satisfactory report issued by the technician at that particular visit can be taken as a blanket statement that the entire project is satisfactory.

REFERENCES

Chen and Associates, Field Manual, revised by Kumar and Associates, 1998.

S. J. Greenfield and C.K. Shen, *Foundations in Problem Soils,* Prentice-Hall, Englewood Cliffs, NJ, 1992.

Byrum C. Lee, Construction Litigation.

17 Legal Aspects

CONTENTS

Suits against professionals are generally categorized under the term "malpractice." The legal standard for determining professional negligence requires proof that a person who is considered by the community to be a professional (as determined by education, training, licensing, and/or experience) has failed in rendering service to meet the standard of ordinary care.

Engineers, as professionals, are held to a standard of ordinary skill in their rendering of service. The services of experts are sought because of their special skills. They have a duty to exercise the ordinary skill and competence of members of their profession and a failure to discharge that duty will subject them to the liability of negligence.

Over the years, earthworks and foundation claims have become considerably more sophisticated in terms of technical arguments, and they are prime issues of dispute resolution. Analyzing and evaluating the facts, in a straightforward and understandable way, and presenting an effective defense against these claims require special skills. The effectiveness of these efforts governs the success or failure of the respective parties.

17.1 LIABILITY CLAIMS

The frequency of claims for design professionals has increased dramatically in the last 40 years. In 1960, there were 12 claims per 100 firms; today, the number increased to more than 50.

17.1.1 Liability by Number

The following data clearly indicate the problem of consulting engineers facing liability claims:

1. The amount that the average consulting engineering firm pays for professional liability insurance has increased by 23% since 1986.
2. Over 5% of firms' annual billings are consumed by insurance premiums. Keep in mind that the net profit of the average consulting firm is between 5% to 10%.
3. The problem of increasing premiums is compensated for by higher deductibles and less coverage.
4. The number of firms "going bare" has hit a new record. More than 23% are uninsured, up from 19% in 1986.
5. Engineering firms turned down 69% more work in 1996 than in 1995 because of potential liability exposure.
6. Consulting engineering firms have learned that promising new approaches to engineering solutions can become lightning rods for lawsuits if not supported by a long record of results.

The trend of liability claims is such that:

1. The larger the firm, the greater the exposures.
2. Suits are usually filed against the most competent, most successful firms. The incompetent and the one-man firm are seldom involved in litigation.
3. More suits are filed against firms with large insurance coverage. This is what is commonly known as the "deep pocket" phenomenon.

It has been suggested that suits against engineers will keep the engineers on their toes so that it is possible to screen out the incompetent engineers and keep the best in the profession. In fact, engineers today exercise much more care in avoiding and preventing liability problems than two decades ago. Most suits against engineers are not based on technical errors but on minor details involving inconsistencies from which attorneys create big issues.

17.1.2 ASFE

Prior to 1970, geotechnical engineers ranked number one among design professionals as to the frequency of being sued. Owners found that it was easy to blame foundation movement on structural distress. At the same time, it was difficult to prove that the distress is not foundation-related. In fact, all insurance companies refused to insure geotechnical engineers because of the high risk.

Finally, the Association of Soil and Foundation Engineers, Inc. (ASFE) was founded. The ASFE devoted a substantial portion of its time to development of a liability loss prevention program. Geotechnical engineers found that the in-house quality control programs such as those listed below can certainly discourage suits.

1. **Failure to supervise inexperienced employees** — A major portion of minor error is the work of inexperienced employees. Most geotechnical claims involve field control. Field supervisors sign reports that relieve the responsibility of the contractors.
2. **Design changes** — Because of the changes requested or required by the owner, the design professional subjected himself to unreasonable time schedules that do not permit adequate review and checking.
3. **Contract specification** — Architects use out-of-date specifications that do not reflect actual project conditions. Contracts that fail to adequately define the duties and responsibilities of the professional and his professional function create an unnecessary and unwarranted exposure to claims.
4. **Certification** — Typical of the documents being used are those that require consulting engineers to certify their findings. "Certify" is a treacherous word. Virtually all professional liability insurers interpret it to be "a promise made in writing." As such, it creates a contractual liability, and therefore it is uninsurable.
5. **Language** — Some of the most severe claims brought against engineering firms have been the result of simple typographical errors that were not caught either by the typist or by the person reviewing the report. Most engineers are aware that one does not "inspect" the projects; one only "examines." By inspection, one virtually takes over the responsibilities of the contractor. In report language, one must be very careful in using such words as "may" or "can," and how often one uses the word "determine" instead of "evaluate."

With the loss prevention program, members of the ASFE are able to drastically reduce liability suits. Today, the high-test percentage of lawsuits is filed against civil engineers (41%), followed by structural (11%), and mechanical (9%) engineers. The insurance cost of geotechnical engineers is even lower. Insurance companies are now soliciting business from the geotechnical engineers.

17.1.3 Potential Plaintiffs

While most professionals, such as doctors, lawyers, or dentists can be sued only by their clients for their failure to properly perform (in rare cases, a third party), the engineer must protect himself and defend his professional conduct against claims by:

The client
A subsequent purchaser
The general contractor
Subcontractors
The contractors' or subcontractors' employees

Governmental or quasi-governmental agencies
Members of the public
Other professional parties to the performance of the work, such as architects, structural engineers, landscape architects, etc
Independent groups such as conservation societies and material or component suppliers

This large group of potential plaintiffs creates a special problem for engineers. For geotechnical engineers, the list of potential plaintiffs can be even longer.

17.2 SCOPE OF LIABILITY SUITS

Almost all liability suits against an engineer start with the familiar sentence that the engineering firm has not met the "standard of care." Before the turn of the century, engineers were considered immune from liability for their errors. By 1900, this immunity from negligence liability was largely lost, and architects and engineers were made to pay damages if they violated their professional duties to a client.

The key confusion as to the term "standard of care" is as follows:

1. What constitutes "ordinary skill and competence"? What is considered "ordinary" to one may not be so ordinary to others.
2. What constitutes "fallacy"? In engineering, the concept of "fallacy" implies designs below the factor of safety. Most of the suits filed today are far less serious in nature.
3. The term "service" should be emphasized. Those who hire engineers are not justified in expecting "infallibility," but they can expect reasonable care and competence. They purchase services, not insurance.

Professional liability is, in fact, an objective standard imposed upon the professional and measured by a reasonably prudent practice for those engineers in similar activities and in the same geographic area. Thus, it is obvious that the standard changes from time to time and from place to place. What is acceptable practice today may not be acceptable tomorrow or may not be acceptable today in another community.

17.2.1 Frivolous Suits

One of the primary concerns of the design professionals in private practice is the debilitating financial drain resulting from lawsuits that have no merit or factual basis. Such unfounded lawsuits are commonly referred to as shotgun suits or frivolous suits, e.g, the plaintiff names anyone and everyone even remotely connected with the occurrence that caused the injury.

After expending a great deal of effort, the engineering community in Colorado succeeded in enacting a bill to limit frivolous suits. However, the presiding judge is the only one who can decide whether the case is frivolous. It has been very rare that the status of frivolous suits has been applied.

One way to combat this type of lawsuit is by counter-claiming against both the attorney and the client. Another way is counter-claiming for abuse of the civil process. Both approaches are discouraged by the attorney, as well as by the insurance company. They prefer to settle the case first. Consulting engineers cannot afford to spend the time and money to fight frivolous suits.

Of course, the best strategy might be for Congress and the state legislators to adopt the English system of forcing losers in civil suits to pay the winner's legal fees plus court fees.

A geotechnical firm made a foundation recommendation for a new jail in Wyoming. After completion, an inmate escaped from his cell and commandeered a pickup. Upon reaching Montana, the inmate shot and killed the pickup owner. The family of the deceased sued the Governor of Wyoming, the jail warden, the architect, the engineer and the geotechnical consulting firm. Of course, this was a frivolous suit. The engineer and especially the geotechnical engineer had nothing to do with the jailbreak, yet it took several months and several thousands of dollars in attorney fees before the suit was dismissed.

In Aspen, Colorado, a developer used a soil report for a site apart from his development without the knowledge of the geotechnical firm. When problems emerged from his development, he sued the consulting engineer for improper soil investigation. Again, in order to dismiss such a ridiculous suit, considerable attorney fees are required to comply with the legal procedures.

A design professional who becomes an undeserving defendant, frequently discovers that the plaintiff is quite willing to drop the suit for a settlement figure of several thousand dollars. Frustrated, angry, and convinced that he is the victim of extortion but recognizing the harsh realities of the cost of the defense, the design professional may accept a settlement offer for purely economic reasons.

Such "legal blackmail," if not put to a stop, actually encourages as much as 35% of lawsuits against engineers.

17.2.2 Contingency Fee

Under the current system, if the person bringing suit wins, the attorney receives a contingency fee of about 33% of the award. If the person bringing the suit loses, the attorney receives nothing.

The argument in favor of this system is that people who otherwise cannot afford a lawyer can readily obtain one. On the other hand, in the case of engineering malpractice, such a system encourages owners to sue upon the discovery of a slight fault in the construction. The owners understand very well that the damage is slight and that it could have resulted from other causes. But his attorney encourages him to sue by saying "What have you got to lose? If you win, you get two thirds of the award; if you lose, you don't have to pay me anything." The greed factor encourages the lawsuit. If the contingency fee system is eliminated or modified, it is believed that more than half of the suits against engineers would not take place. Furthermore, the contingency fee system offers the attorney an incentive to sue for greater awards because larger settlements mean larger fees. It is seen over and over that home owners sue the builder, the architect, and the soil engineer, when such suits would never take place without a contingency fee arrangement.

A Gallup survey indicated that 32% of respondents said lawyers should be compensated through contingency fees; almost twice that number believe that lawyers should receive a fixed fee in advance, regardless of whether the case is won. Others believe that lawyers' fees should be based on the number of hours the lawyers actually work on the case.

17.2.3 Comparative Negligence

As a general rule, when an accident occurs on a construction site, everyone connected with the project is brought into the resulting suit. Under the contributory negligent rule, the plaintiff should not recover damages that they caused themselves by retaining the 50% cutoff for recovery.

In comparative negligence jurisdictions, the plaintiff's own negligence may not bar him from obtaining any recovery at all. The owner, however, is only able to recover a judgment for the percentage of the claim that corresponds to the degree of capability of the defendant for the injuries.

In most cases, the design professional's degree of capability will be found to be relatively small compared to that of the construction contractors and owners. Accordingly, the change from contributory negligence to comparative negligence, in many cases, will result in a smaller judgment against engineers.

It is difficult to assign the percentage of negligence. In most cases, geotechnical engineers are not involved with projects, except for the initial soil reports. And in many cases, the actual construction is essentially different from the preliminary concept. Yet when the geotechnical engineers are brought into the suit, they cannot escape a percentage of negligence. The smallest share is about 5%. For a multimillion-dollar suit, this 5% can be a considerable amount.

17.2.4 Joint and Several Liability

Another major issue is the "joint and several liability" rule. This rule, dating from the 19th century, holds that where several defendants, either in concurrence or independently, cause damage to the plaintiff, the plaintiff can recover entire damage awards from any one defendant, usually the one with the "deep pockets."

An example of comparative negligence is one with the general contractor's share at 80%, the structural engineer at 15%, and the geotechnical engineer at 5%, but the preliminary outcome changed. It turned out that the contractor declared bankruptcy, the structural engineer carried no insurance, and the geotechnical engineer carried high insurance, so the 5% firm had to pay for the total damage. Of course this is an extreme case, but it happens. Fortunately, many states under "tort reform" have written off this rule. A party is now required to pay its share of fault, as assigned by the jury.

17.2.5 Punitive Damages

Punitive damages in a pain and suffering award have become quite common in suits against engineers; especially in the case where the actual damage is small, such as a residential house, the amount of punitive damage can be 10 times higher than the repair cost.

People bringing suits may ask for punitive damages in addition to compensation for repair work and loss of service. Punitive damages under the heading of "mental distress" or "inconvenience" can amount to millions of dollars. Generally, people seldom bother dragging a minor case into court, but the prospect of thousands or millions of dollars in punitive damages might encourage them to do so.

Judgments for punitive damages are a major cause of a high increase in court cases. The large amount that may be paid under these categories caused some lawsuits to go to trial that otherwise might have been settled voluntarily.

A basement house was flooded from the rise of perched water. The builder installed a drain system with a sump pump and repaired the damage with a new carpet. The housewife demanded punitive damages. She brought her three children, aged 3 to 10, into court. During the testimony, she fainted, and her three children rushed to the witness box to console her. The entire drama impressed the jury that the full punitive request was awarded.

17.3 CHANGE OF CONDITIONS

The change of conditions clause, or as it is now often called, the differing site conditions clause, is one of the most significant risk allocation clauses found in a construction contract. Without the clause, the contractor bears most of the financial risk associated with the encountering of onerous job-site conditions unforeseeable at the time of bidding. With the clause, the owner accepts that risk.

On the other hand, for a highly competitive bidding contract, some contractors may intentionally lower the bid in order to get the contract. They expect to recover any loss through arguing about the site conditions; with the change of conditions awarded to them, they may gain financially.

There is no limit to the factual situations giving rise to a changed condition. Typical of the conditions that have been found to qualify as changed conditions that concern geotechnical engineers are:

1. The presence of rock or boulders in an excavation area where none or few were shown on the drill log
2. The encountering of rock or boulders in materially greater quantities or at different elevations than indicated in the drill log available to bidders
3. The encountering of ground water at higher elevations or in quantities in excess of those indicated in the data furnished to the bidders
4. The difficulty in adopting a drilled pier system as recommended
5. The elevation of bedrock
6. The necessity of using casings to complete the drilled piers

Most engineering contracts are provided with clauses that stipulate the geotechnical report attached is for the use of the bidder as reference. The bidder should conduct his or her own investigation to verify the accuracy of the document. However, in a court of law, the owner cannot avoid the responsibility of the information provided to the bidders.

A change of conditions as interpreted in the court can best be illustrated by a case that took place several years ago. A bridge foundation contract was awarded to a pier drilling company in Arizona. When the contractor encountered unusual difficulties, he asked for compensation under the "change of conditions" clause. The main arguments of the case are as discussed below.

17.3.1 Drill Logs

As stated by the Bureau of Reclamation, a log is a written record of the data concerning materials and conditions encountered in individual test holes. It provides the fundamental facts on which all subsequent conclusions are based, such as the need for additional exploration or testing, feasibility of the site, design treatment required, cost of construction, method of construction, and evaluation of structural performance. A log may represent pertinent and important information that is used over a period of years; it may be needed to delineate accurately a change of conditions with the passage of time; it may form an important part of contract documents; and it may be required as basic evidence in a court of law in case of a dispute. Each log, therefore, should be factual, accurate, clear, and complete. It should not be misleading. Discussion of a drill log is given in Chapter 4.

It is therefore important that the engineer in charge of logging of the drill holes should be observant and experienced. Unfortunately, the man in charge of drilling in the following case had little knowledge in logging the test holes.

The field log provided little information under the soil description column. The subsoil condition of the bridge site was summarized in the soil report, in which the report stated:

"...All ten borings encountered very loose to dense silty sands with occasional cobbles overlying an interbedded claystone and sandstone with occasional layers of unconsolidated clean sands..." The penetration resistance test as shown on the field log indicated that the lowest value was 7 and the highest value was 73, with an average value of 40 for the more than 100 tests. Based on the standard classification method as described in Chapter 5, the density of the soil should have been described as dense to very dense, instead of very loose to dense. During drilled pier installation, the driller found that the actual field conditions were quite different from those stated in the report. Some of the major differences were as follows:

1. "Occasional" is defined as "coming irregularly, occurring from time to time." Large amounts of cobbles were encountered in each pier hole, some occupying as much as 60% of the soil. Therefore, the word "occasional" was misused and caused misinterpretation.
2. During drilling of the test holes, on one occasion, it took 3 days of drilling to complete 6 ft of the upper subsoils. The field engineer should have

noticed the difficulty and explained the unusual condition to warn the bidder. Instead, the drill logs still use the word "occasional cobble."

3. In the drill log, the presence of "boulders" was never mentioned. In fact, boulders constituted a major part of the subsoil and were probably the main reason for the difficulty in the installation of the drilled piers.

Based on the information furnished from the field log, the geotechnical engineers recommended the use of drilled piers for the bridge foundation. Actually, had the geotechnical engineer been aware of the actual subsoil conditions at the bridge site, he may have considered the use of a footing foundation instead of drilled piers at a considerable saving in construction costs.

The above case fully illustrates the justification for the contractor demanding additional compensation on the ground of a "change of conditions." On the other hand, before entering the bid, an experienced contractor should visit the site and, upon seeing the exposed large boulders, he would question the accuracy of the drill log.

17.3.2 Depth to Bedrock

The drilled pier foundation depth to bedrock is an important issue affecting the cost of the project. For most geotechnical engineers, depth to bedrock appeared to be an easy fact to determine. However, where bedrock consists essentially of claystone, it is sometimes difficult to differentiate the upper claystone from stiff clay. Geotechnical engineers arbitrarily classify claystone with a penetration resistance exceeding 20 as bedrock. Pier drillers tend to drill the piers deeper than indicated in the documents and charge the excess penetration as a "change of conditions." In order to avoid confusion, it is therefore important to point out such possibilities clearly in the geotechnical report.

On a river crossing where a drill rig cannot be used, the usual alternative is the use of a geophysical method to determine the elevation of bedrock. Referring to the above case in Arizona, a geometric 12-channel seismograph device was used to determine the depth to bedrock in the bridge crossing. The investigation was conducted by university students.

The report stated "… to minimize the error in depth, depth is determined as plus or minus 4 feet." During construction, it was found that the report underestimated the depth of bedrock by as much as 20 ft. Upon further investigation, it was found that the geophysical survey conducted by the students was not accurate and did not use up-to-date technology. The geological report was presented to the court as major evidence in asking cost overrun.

17.3.3 Specifications

Specifications are a part of the contract document to which both the owner and the contractor must adhere. Soil classification systems used by various institutes are given in Chapter 4. For the definitions of "cobble" and "boulder," all systems have a similar definition. The unified soil classification gives:

Cobble — A rock fragment usually rounded or semi-rounded with an average dimension between 3 and 12 in.
Boulder — A rock fragment usually rounded by weathering or abrasion, with an average dimension of 12 in. or more.

The above definition uses the term "average" not the "least." It is confusing as to whether the term "average" refers to three- or only two-dimensional measurements.

ASTM definitions are more straightforward. They state as follows:

Cobble — Retained on 3 in. and passing 12 in. sieve.
Boulder — Not passing 12 in. sieve.

The above definition of cobble left an open discussion of classification. It is possible that a rock fragment is able to pass a 12-in. sieve, but with a length of several feet, and still be classified as cobble.

In the case of foundation drilling for the bridge structure, the contractor encountered numerous large boulders in every drill hole and very difficult drilling conditions. According to the owner, such huge rocks are still classified as cobble. The seemingly inconsequential definition can mean a million-dollar lawsuit. Figure 17.1 indicates the boulder in the drilled pier hole.

17.4 EXPERT WITNESSES

In most legal cases, expert witnesses are required. This is especially true when engineering practice is involved. The court must establish that a case exists and must prove that the defendant is the party responsible for the said damage and should be held accountable. It is obvious that most attorneys have little knowledge of engineering and must rely on experts to analyze the problem. During the trial, the judge as well as the jury must rely on experts to explain the problem in layman's language.

Clearly, in a complicated engineering project, it is a battle between experts. Unfortunately, most engineers have little or no experience in serving as expert witnesses. When they first appear in court or at deposition, they often find the experience far different from any they have ever known. Their prior experience in giving technical reports does not prepare them for the possible jolt their egos may undergo while appearing as expert witnesses.

17.4.1 Standard of Care

Before agreeing to serve, one must ask if there is an understanding of prevailing standard of care as it applies to the circumstances at issue. One must know what the average engineer might reasonably have done under similar conditions before an opinion can be formed about what, in fact, was done. One must think in terms of ordinary skill and care; if not, one is probably the wrong expert. The key question a plaintiff's attorney will ask is as follows:

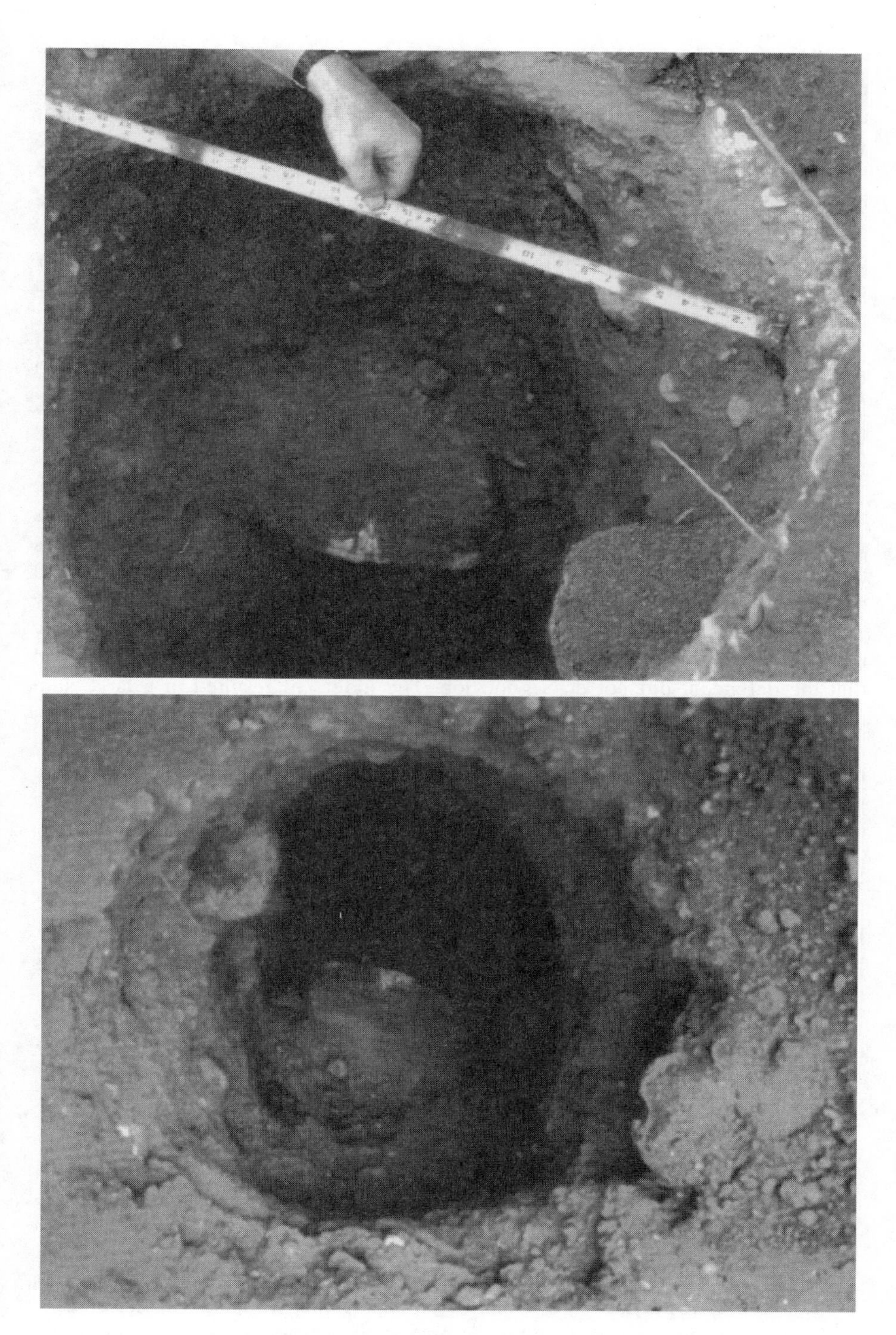

FIGURE 17.1 Boulder in the drilled pier hole.

"Has the engineer in his/her work employed that degree of knowledge ordinarily possessed by members of that profession, and to perform faithfully and diligently any service undertaken as an engineer in the manner a reasonably careful engineer would do under the same or similar circumstance?"

This is a lengthy question. In fact the attorney is asking not what would have been done, but if what has been done was reasonable at that time and in those circumstances. If the answer is "no," the defendant is doomed. If the answer is "yes," the case will continue, and the defendant has a good chance. Remember, one may not agree with his or her conclusion and presentation, but he or she still has exercised the standard of care.

The best solution, whatever that might be, is not the issue. The issue is usual and customary care. One must differentiate between reasonable care and substandard performance. This puts the defendant in a favorable position. As an expert witness, it can put someone on the spot, because serving as an expert can be a challenging task. It is not to be taken without a good deal of careful consideration.

A silo designed by a reputable geotechnical consultant had settled much more than the soil report indicated. During the trial, the plaintiff engaged a world-famous geotechnical engineer as an expert. He reviewed the report, analyzed the approaches the geotechnical engineering consulting firm used to reach the settlement figure, and found many areas where he would have done it differently. He stood before the jury and delivered something like a Terzaghi lecture. Neither the judge nor the jury understood what he was talking about. Still, with his reputation, the court decided that the defendant did not do a good job. The judgment resulted in a large sum awarded to the plaintiff.

It is therefore important to understand that a consultant gives opinion and advice but does not guarantee performance. It is up to the insurance company to guarantee performance. For average projects, the owners pay the consultants only a fraction of what they pay for the insurance.

17.4.2 Ethics and Egos

Legally, an expert is a person who, because of technical training and experience, possesses special knowledge or skill in a particular field that an average person does not possess.

Unfortunately, the selection of an expert by the attorney has been abused. Some define "expert" as any person of average knowledge who is more than 50 miles away.

After being named as an expert, the consultant must be prepared to perform. Some important considerations that must be understood and addressed are:

1. **Agree to serve** — Agree to serve only if one is convinced that the interests of both the profession and the public compel one to become involved. One should refuse to serve if one thinks that there is no case, no matter what amount of money is offered. At the same time, if at all possible, one should help one's fellow professionals in seeking justice.
2. **Area of expertise** — One can be an expert only in one particular area. It is sometimes very tempting to give an answer and express an opinion

on something apart from one's expertise. There have been countless examples of irresponsible comments by structural engineers, some of whom even discredit a geotechnical engineer's finding. An experienced attorney will object and throw out such testimony, but frequently such comments will be regarded by the jury. Such an important ethical issue should not be overlooked by fellow engineers.

3. **20/20 hindsight** — The constraints imposed on the original design must be understood. There are very few problems that could not be prevented if approached with the 20/20 hindsight one brings to the dispute. Resist the temptation to make judgments based on one's high standards. One may be the best engineer, but one is not always present in the courtroom to explain how much more effectively the design should have been prepared.
4. **Resist pressure** — Be prepared to resist pressure to arrive at conclusions desired by those who are going to pay the consultation fee. One has the obligation to meet the expectations of the attorney who is one's client. One's attorney will at times exert pressure to accuse the opposing engineer of failing to comply with the standard of care, unless there is an inexcusable error or statement.
5. **Tactics** — When testifying in court, it is important that the witness give the jury a good impression. One must dress properly, and speak clearly and with authority. One must be positive and not retract one's answers. Experienced attorneys will ask a series of unrelated questions leading to a key question. The tactic is to confuse the witness, who will sometimes give the wrong answer. On the other hand, an experienced witness is able to upset the interrogating attorney and induce him or her to ask stupid questions. Do not hesitate to ask the attorney to repeat or rephrase the questions. To ease the trial tension, it is sometimes advisable to inject humor into your answer.
6. **Use of exhibit** — The use of exhibits is an important tactic for the expert. Judges and juries can understand the case more clearly if simple graphs or charts can be presented. The use of models prepared in advance of the trial can be very impressive. When a small slab movement or a beam crack is enlarged ten times, judge and juries will be impressed. Poorly prepared models sometimes may be thrown out by the judge.
7. **Deposition** — In deposition, the expert witness will be allowed to express his opinion on the cases almost freely. The opposition will object but will allow the witness to carry on. However, the witness should understand that whatever he testifies is on the record and can be used to discredit him later at the trial. More than 80% of litigation against engineers is settled out of court after lengthy depositions.

17.5 THE OUTLOOK

There are about 700,000 lawyers in the U.S., more than in the rest of the world combined. For every dollar awarded in litigation, 67 cents goes to the attorney and 33 cents goes to the injured party.

Consulting engineering firms that buy more professional liability insurance and make horrendous investments in non-chargeable staff time for litigation are clearly on the road to bankruptcy. In the past 30 years, the author has participated in over 100 legal cases, including court appearances, arbitration, and depositions. The more one participates in legal cases, the more frustrated, angry, and desperate one becomes. One often wonders whether this is the way justice is served.

Our legal system must be preserved, but more tort reform must be made so that the design professionals can freely express their ideas where the client's money can be saved by intelligent design, and where new concepts can be created without the fear of being sued. Our technology cannot be improved to compete with foreign markets if the shadow of lawsuit is kept hanging over our heads.

It is obvious that something must be done in the following areas, on both federal and state levels:

1. A cap on liability awards, especially on punitive damages
2. Limitation on contingency fee arrangements
3. Providing a fair statute of limitation
4. Stiff penalties for frivolous suits
5. Settlement of disputes quickly, and avoidance of lengthy trials
6. Reeducation of the judges for basic principles of faults and negligence
7. Elimination of the outdated "joint and several liability" law
8. Encouragement of the use of "arbitration," both binding and non-binding, for the settlement of disputes
9. Further investigation of the use of "mediation" for dispute settlement
10. Imposition of "Certification of Merit" before the court will take the case

With the professional societies working hand-in-hand with the legal profession, it is expected that a better climate in this muddy field can be achieved.

REFERENCES

ASFE, Alternate Dispute Resolution for the Construction Industry, Silver Spring, MD, 1989.
R.F. Cushman, *Avoiding Liability in Architecture Design and Construction,* John Wiley & Sons, New York, 1983.
R.F. Cushman, *Differing Site Condition Claims,* John Wiley & Sons, New York, 1992.
H.W. Nasmith, *Suit is a Four Letter Word,* Bitech Publishers Ltd., Vancouver, Canada, 1986.
R.C. Vaughn, *Legal Aspects of Engineering,* Kendall/Hunt Publishing, Dubuque, IO, 1977.

18 Report Writing

CONTENTS

The final and most important phase of an investigation is the report. Most clients are not interested in how the consultant conducts the field investigation, the laboratory testing, or the mathematical derivation; they are interested only in the recommendations offered by the consultant on their projects. Unfortunately, most engineers have no training in the art of report writing. In many universities, courses in English have long been replaced by more popular subjects such as finite element analysis.

18.1 TYPE OF REPORT

Geotechnical consultants should understand the type of business in which their clients are engaged and their reasons for requiring the soil report. For developers, the purpose of obtaining a soil report is mainly for the fulfillment of the regulatory agencies such as lending agencies or home warrantee associations. They tend to shop for the consultant who provides the service for the minimal fee. When problems arise, they point their fingers at the consultants. For such clients, it is important to state clearly in the report what to do and what not to do. Theoretical approaches leading to the conclusions are neither helpful nor necessary. When the report is directed to companies with an engineering staff or to an academic institution, it is important to qualify all statements.

18.1.1 Letter of Transmittal

In the letter of transmittal, it is desirable to clearly state the intent of the assessment. In addition, the following should be included:

1. Name of firm must be correct, including firm's practice
2. Name and title of addresses
3. Reference project title and architect/engineer job number and authority
4. Names of those on the distribution list

18.1.2 Investigation Proposal

The subject heading should clearly indicate the type of study, whether preliminary, feasibility, technical, etc. In addition, the following should be included when necessary:

1. **Project Description** — Items such as type of construction, loads, grade changes, project location, and/or special requirements should be included.
2. **Purpose of Study** — A clear statement is required. Be specific and discuss limitations.
3. **Scope of Service** — A clear and specific statement of items of work to be performed. This may include one or several of the following:
   a. *Exploration*
      Type and equipment
      Number
      Estimated depth
      Layout and elevations
      Observation
      In-house subcontract
      Special equipment of in situ tests
   b. *Laboratory testing*
      Index properties
      Physical properties
   c. *Analyses*
      Settlement
      Stability
      Swell
      Other
4. **Fees** — Method of charging, estimate of total cost (for upset, or lump sum as appropriate), and billing procedures.
5. **Conditions** — This includes standard clauses used in engineering contracts plus other conditions as desired/required.

a. Limitation of liability
b. Warranty
c. Definition of responsibility
d. Provision by client
   Project description, including loads, grading, etc.
   Right of entry to do exploration
   Location of subsurface utilities
   Survey reference
   Special form of invoicing, report format, design, and construction standard to be followed

6. **Performance Schedule** — Time required to complete the study
7. **Authorization or Acceptance**

18.1.3 Preliminary Report

The types of reports most frequently required from the geotechnical consultant are the preliminary report and the final report. The preliminary report is often required by the clients for the following reasons:

1. When the client is not certain whether the site is suitable for the development
2. When the client wants to know in which area to erect the structure on a large plot of land
3. When the client wants to know whether there are any unusual subsoil conditions that require special treatment
4. When the client wants to know the most preferable location to place his or her structure

The preliminary report prepared by the geotechnical consultant is usually very brief. The field investigation usually consists of no more than a couple of test holes to determine the depth to bedrock and the level of groundwater. Recommendation about the type of foundation system may not be given and the design pressure may not be determined. The structural engineer cannot use the preliminary report to design the foundation.

Unfortunately, some clients may consider the preliminary report final and do not conduct further soil tests in order to save some consulting costs. Therefore, it is essential that the geotechnical engineer should strongly stress that the preliminary report cannot be considered final and the information provided in the preliminary report cannot be used for design purposes.

18.2 GEOTECHNICAL REPORT

The final geotechnical report is not only used to fulfill the contractual obligation with the client, it can also be used as a permanent document. It may be used as evidence in the court. It may be refereed by the academicians to dispute a certain theory. It may be pointed out by the professionals as inadequate. In some extreme cases, the developer may claim that the soil report devalues the property and threaten legal action. The punch lists for a complete geotechnical report are listed as follows:

18.2.1 Geological Description

Surface geology, emphasis on glacial, formation of moraines, shorelines, benches, eskers, and other visible features
Effects of glacial lakes, wind, and other erosion
Depth and character of overburden
Effects of weathering, dip, strike, and thickness of drift
Type of bedrock
Incidence of well water, artesian, static level, probable volume, and contamination
Presence of gas — methane

18.2.2 Drainage

Limits of watershed
Principal drainage paths and direction of drainage
Seasonal conditions, erosion
Fifty or one hundred-year storm precipitation
Surface drainage

18.2.3 Site Conditions

Locate site with respect to adjacent or nearby street and point of compass.
Describe community by township, county, city, or state.
List and describe topography of site, together with notations of trees, drains, ditches, or other natural features.
Note existing structures and other man-made installations visible to the naked eye.
Refer ground surface elevations to known benchmark, or establish own benchmark.
If the site is reasonably uniform, describe the typical soil profile.
Describe fill deposits in general terms only and qualify depths indicated on the logs.

18.2.4 Boring Conditions

List authorization of boring locations by architect's or engineer's print, owner's sketch, or by owner's/architect's/engineer's/contractor's representative at site.
Find static number and depths of borings.
Refer to location of borings on print or sketch and note origin of drawings.

Ascertain economic values of overburden or rock.
Find incidence of well water, artesian, static level, probable volume, contamination.
Find depth of frost penetration.
Describe the character of rock encountered and method of identification (drill water, chips or cuttings, drive sample of soft rock, cores, test pits, etc.).
Note weathering, jointing, planes, honeycombing, hardness, coloring, porosity, recovery percentage.
Note any water loss and drilling conditions.
Describe the range and depths or elevations at which water was first encountered, also range of depths after a period of hours or 24 hours.
Comment as to whether water is perched, related to seams of silt or sand, seasonal, artesian, apparent head, or appears to be contaminated.
Note presence of gas, method of detection, apparent elevation of origin, volume or pressure, and identification.
Describe methods used to fill drill holes to prevent inflow or surface water, or the interconnection of aquifers, or causes of accidents.
Describe grouting procedures and depths to stop artesian flow or escape of gas.

18.2.5 Proposed Construction

Describe information provided, with respect to size and character of structure.
Determine live and dead loads provided for interior and exterior columns and also column spacing.
Will a basement be required? Will there be tunnels, machine pits, elevator pits, or other substructures?
Has information been provided with respect to special installations such as overhead cranes, compressors, vibrating machines, or impact load?
Is the settlement of the building or machine critical?
State information on finished floor grades, outside grades, and whether filling is required.
Emphasize that analysis and recommendations in report are based on the information provided and are critical.

18.2.6 Discussion of Foundation Types

Discuss the merits of only those types of foundations relative to the particular project. Consider all types of foundations, such as:

1. **Spread Footings** — Discuss the following items:
   Strip or continuous footings, column footings, basement footings
   Placement of reinforcing
   Possibility of settlement, both total and differential
   Time factors

Economies of excavation and backfill
The placing of spread footing on compacted fill
The effect of the groundwater table
Design soil pressures for the various footings at various depths and under various conditions of settlement

2. **Mat or Raft Foundations** — Discuss the following items:
   Size and probable thickness necessary for the reinforcing
   Differential settlement from outside to center
   Tipping due to unequal consolidation beneath mat or uneven loading
   Depth of pressure bulb
   Difficulty of installation below floor or foundation drains or utilities
   Effects of filling site adjacent to mat
   Economical comparison of the system with other foundation systems
3. **Driven Piles** — Discuss the following items:
   Use of various types of piles, such as pipe, wood, composite, step-tapered, monotube, H-section, or concrete
   The necessity for mandrel pre-boring, or the use of batter piles. Hammer size, cushion block
   Minimum penetration or resistance
   The effect of overdriving
   Bearing capacities of friction and end-bearing piles
   The use of single piles versus cluster piles
   Negative skin friction
   Resistance to lateral forces
   Vibration, displacement, and heaving
   Redriving effects on nearby structures
   Plumpness requirements
   Concreting schedule and verification of location
   Pile corrosion, treated timber, and marine deterioration
   Test piles, number and procedure of testing and interpretation of test results
   Testing, inspection, and pay items
4. **Drilled Piers** — Discuss the following items:
   Advantages and disadvantages on the use of drilled piers
   Speed of installation, minimum depths and shaft diameters
   Bearing elevation
   Bearing capacity
   Bell piers
   Friction on shaft for support of load
   Negative skin friction
   Casing
   Expansive soils, dead load pressure
   Sealing against water, hydrostatic pressure
   Pulling casing
   Concreting precautions against voids
   Cleaning

Testing and inspection
Verification on volumes for quantities

5. **Combined system** — Discuss the following items:
   Different foundation systems
   Addition to existing building
   Differential settlement
   Magnitude and rates of settlement
   Construction joints and expansion joints
   Waterproofing of roof joints
   Effects of new foundations on adjacent existing structures
   Necessity of underpinning
   Underpinning methods and precautions
   Necessity of shoring and bracing
   Estimates of jacking resistance available.

18.2.7 Groundwater Identification

The following items should be included as deemed necessary:

1. **Type of water** — Discuss the following:
   Surface, perched, artesian
   Natural springs
   Seasonal variations in levels and volumes
2. **Control** — Discuss the following:
   Surface runoff
   Ditches — permanent or temporary
   Construction of sumps and pumping
3. **Disposal of water** — Discuss the following:
   Deep wells, size, and spacing
   Estimate of phreatic surface and rate of drop in water level
   Dewatering with wellpoints
   Permeability of soils
   Effects of lowering water table on adjacent structures
4. **Type of drain** — Discuss the following:
   Type of drain and surrounding filter media
   Interior sumps and hydrostatic pressure relief valves

18.2.8 Site Development

The following topics should be included in the discussion as deemed necessary:

1. **Cut and fill requirements**
   Balancing requirements
   Suitability of available borrow
2. **Select fill requirements**
   Preparation of subgrade
   Stripping, grubbing, and clearing

3. **Proof rolling or compacting**
   Suitability of equipment
   Moisture control
   Compaction requirements
4. **Preparation of subgrade for floor slab**
   Necessity for reinforcing
   Expansion joints and control joints at wall and foundation
5. **Preparation of subgrade for pavements**
   Evaluation of subgrade by plate-bearing tests for modules of subgrade reaction, for rigid pavement
   Evaluation of subgrade for flexible pavement by California Bearing Ratio method
   Review of pavement costs, performance and durability
   Effect of frost, traffic load, and densities
6. **Stability requirements for asphalt concrete**
   Concrete strengths and air entrainment
7. **Requirements for underdrains subdrains**
   Edge drains and around catch basins

18.2.9 Failure Investigation

The following is a checklist for distress investigation of cracked structures:

Require personal inspection by engineer.
Complete history of structure from planning stage to the time of investigation.
Obtain evaluation and advice from a structural engineer.
Vertical controls and monitoring of settlement must be made by a registered land surveyor.
Determine slab-related or foundation-related problem.
Provide a detailed description of results of investigation and evaluation.
Report limitation.
Since most failure investigations are legally involved, precaution should be exercised in the choice of words.
Pavement failure investigations are less critical but need complete testing and evaluation program.
Appendix must be complete, but avoid the inclusion of unrelated matters.

18.2.10 Inspection and Testing

Virtually all projects require and benefit from a complete testing and inspection program. The report should include the following:

Review the necessity for full-time inspection and control, versus part-time.
Stipulate type of inspection required and whether an engineer or experienced technician is required.

List testing requirements, and types of reports to be furnished.
Stress the avoidance of delays in construction, and the availability of records to settle disputes.

18.2.11 Conclusions

Conclusions can be better placed at the front of the report, or at the end, with these precautions:

Avoid long descriptions. Try to be short and precise.
Indicate opinion on the best selection of a foundation system.
Note bearing capacities.
Report presence of expansive soils.

18.3 ENGINEERING USE OF WORDS

Many words used by engineers have very special and limited meanings. Because some of them are especially susceptible to misinterpretation or are difficult to explain to a layman or jury, it would be wise to use alternative wording to describe that particular activity.

The following listings are recommended by the Association of Soil and Foundation Engineers (ASFE).

18.3.1 Engineering Jargon

	Engineering Jargon	Preferable Word (or Words)
1.	Approve	Review
2.	Certificate (after grading)	Memorandum
3.	Control (the job)	Control tests, compaction tests
4.	Or equal	Or equivalent, give guidelines
5.	Essential (it is)	Considered, advised
6.	Examination	Observation, reviewing, study, evaluate, look over (the job)
7.	Inspection	Observation, review, study, look over, take density tests
8.	Ensure (to be sure)	So that
9.	Investigation (soil)	Exploration, reconnaissance
10.	Necessary (it is)	Consider, advise, study, evaluate, observe, review
11.	Required	Considered, advised
12.	Supervise	Observe, review, look over (the job), guide, guidance
13.	Assume (to ensure)	So

REFERENCE

The content of this chapter was taken from the documents published by the Association of Soil and Foundation Engineers Inc. (ASFE).

To be a Geotechnical Engineer

Engineers should have experience in all phases of soil.
Engineers should have court experience.
Engineers should have construction experience.
Engineers should have basic structural knowledge.
Engineers should not ignore geological impact.
Engineers should study existing reports.
Engineers should not compromise.
Engineers should not speak ill of fellow professionals
Engineers should go through continuing education.

Index

A

B

C

D

E

F

G

H

I

J

L

T

U

V

W